The "Maintenance Insanity" Cure: Practical Solutions to Improve Maintenance Work

Roger D. Lee

INDUSTRIAL PRESS, INC.

Industrial Press, Inc.
32 Haviland Street, Suite 3
South Norwalk, Connecticut 06854
Phone: 203-956-5593
Toll-Free in USA: 888-528-7852
Email: info@industrialpress.com

Author: Roger D. Lee
Title: The "Maintenance Insanity" Cure: Practical Solutions to Improve Maintenance Work
Library of Congress Control Number: 2017953515

ISBN (print): 978-0-8311-3624-6
ISBN (ePUB): 978-0-8311-9468-0
ISBN (eMOBI): 978-0-8311-9469-7
ISBN (ePDF): 978-0-8311-9467-3

Editorial Director: Judy Bass
Copy Editor: Judy Duguid
Compositor: Patricia Wallenburg, TypeWriting
Cover Designer: Janet Romano-Murray

books.industrialpress.com
ebooks.industrialpress.com

This book is dedicated to all the people I have worked with to solve the problems described in each chapter. May the knowledge gained be of benefit to others facing similar challenges. There is light at the end of the tunnel.

Special thanks go to my wife, Gina, for being understanding and supportive of all the travel and time away from home required by this work.

Contents

Foreword

I started my efforts on maintenance improvement when work orders were hand written on a three-part carbonless form. They were kept in folders in a desk drawer and the only time there was a list made was for a shutdown. We only needed one folder because nobody created them unless they wanted us to work on them that day. In fact,we didn't need a folder at all because Mack would just pile them on his desk and when the pile got too high, he would pull the waste basket up beside the desk and rake them into it. The mechanics were oilfield hands who thought the solution to a problem was a bigger hammer or a cheaper pipe. I was the start of an experiment to see if having an engineer in maintenance would help.

I started my journey in reliability and added planning and scheduling when I was put in a supervisory role. I went off to a week-long school taught by the "Master." The main thing that stuck with me is that he would not sign on to help a company if the president was not willing to approve any emergency job that would disrupt the scheduled work. Roger's book will make it clear how important this discipline is. My company was a large chemical company and they spent freely on development of employees. We learned directly from Deming and Aubrey Daniels as well as other innovators.

We added engineers in maintenance and created a central reliability group. This was when I became acquainted with Roger Lee. He led this group and I saw it as an asset where other managers saw it as interference. Needless to say I got the best and most help from this group and Roger and I, along with many others developed a maintenance program that had such a noticeable impact at our plant site that the company adopted and spread our processes through the company's other plant sites. Our company eventually established a new business of marketing our maintenance strategies to other companies for a fee based on verifiable savings. Roger moved to the central offices and led this effort.

Roger was a sincere and dedicated student and innovator in our journey. I appreciate the fact that he has documented our efforts and learnings to help others create a better way of handling maintenance. He knows what he is talking about because he has been there and done it. He is not just writing about what he saw or heard about. This was not an easy journey. The people who operated the plant had to cooperate and do some things that were not comfortable for them and did not make sense from their point of view. They had to share some control and we had some tough characters (who acted more like kings in charge of their kingdom). No one likes change. And with changes like those things can get worse before they get better. The real challenge is for the change agent to stay the course when almost everyone is saying that things aren't working. But when it does start working and the improvements in time and reduced costs are so evident, even the kings want to know how you are doing it.

One of the greatest myths in maintenance is that improved reliability increases maintenance costs. On the contrary—done properly, as the strategies in this book teach, reliability and significant cost reduction always go hand in hand.

I encourage you to find and use the strategies in this book that will best fit your needs. It took us about 20 years to develop and implement these strategies and I would say that no organization effectively uses all of them. This isn't a step-by-step approach to improving your efforts, but rather a smorgasbord of opportunities for improvement that can be implemented one at a time based on your needs.

Richard Rossow
September, 2017

In Brief: About This Book

This book will focus on leveraging the wisdom of experience to share learnings gained and help you change work processes to yield ongoing developmental advantages.

Techniques will be shared to allow you to tap into the available resources contained in your hourly workers. Solutions address organizational structure, assessments of present conditions, gap analysis with gap closure plans, behavior/result reinforcement programs, cultural change processes, and work process development. *The "Maintenance Insanity" Cure* will help develop and drive your new vision to become reality.

Introduction

The "Maintenance Insanity" Cure provides the answers to address the definition of insanity (a definition frequently credited to Einstein) as doing the same thing over and over again and expecting different results. The book contains numerous maintenance solutions from real-life applications that will improve the future of your organization and help you to achieve success in productivity and cost control. Techniques to break the cycle of repeating the same things over and over again will guide you to a new reality. Many practical tools and files described within this book are available for your use at www.maintenanceinsanity.com.

How can we take the mystery out of maintenance? What is so difficult about doing maintenance? The simple answer is "making everyone happy." With tight cost control and limited resources, maintenance is now being asked to do more with less than at any other time in the past. Ask the people in any maintenance organization what they need, and the single most popular answer is more people. Using the same old outdated processes and throwing more resources at the problem will only result in throwing more money at it. Plus, it is getting harder to find good, skilled resources. One of my mentors (Richard Rossow) shared with me the following saying, which he learned from his stepfather, who said it described his World War II experiences. It seems to fit for most of the present-day maintenance organizations. Known as the "unwilling" motto, it says: "We the unwilling, led by the unknowing, are doing the impossible for the ungrateful. We have done so much for so long with so little we are now qualified to do anything with nothing." Achieving optimum maintenance and reliability for your facility has to become a true partner relationship with operations and all support resources. Our success begins when we realize that we are all in this together.

I am not asking you to change to add more work to your already overloaded day. I am suggesting that you change to improve how you do your work. We will help you determine your present state and then progress to a benchmark application at the end of

The "Maintenance Insanity" Cure that offers one possible option for your maintenance and reliability processes no matter the size of your facility. Please allow yourself to be open to the potential described in each solution.

To make you feel better about your present situation, we start with a couple of extreme case studies and a snapshot of "a-day-in-the-life" of a typical mechanic trying to do his or her job in a reactive environment. We next address moving from reactive to proactive work processes with detailed steps to get started. Planning and scheduling start with a day's and week's look ahead for beginners and then expands to share different options. Communication is highlighted by a three-legged stool analogy to show the true value of equal partnerships. We touch on cost analysis with a six-bucket process to incorporate all costs including vendors and contractors. A simple tool is provided for tapping into your workforce using positive reinforcement techniques. We will explain ways to do self-assessments and how to get others involved in deep-dive comparisons between sites. Benchmarked-level descriptions are provided to give you a reference point for your site. For a mature culture, a method to perform sales and operations planning is shared to help with forecasting and meeting organizational and customer needs. Specialty topics like handling the transition for an acquisition or merger are explained, and shutdown and turnaround processes are shared since they are different from day-to-day planning and scheduling. The final chapter gives a glimpse at the future for maintenance where operations and maintenance are fully linked with very successful results. Following these examples will allow you to break out of your "insanity" cycle and move into the future you have always dreamed of having.

Acknowledgments

Big thanks go to Jerry Wilson, a longtime friend, who wrote me a letter of introduction to Terrence O'Hanlon at ReliabilityWeb. Terry accepted my article that became the starting point for *The Maintenance Insanity Cure*. Terry connected me with Sean Flack, who helped me publish "A Glimpse at the Future" in their *Uptime Magazine*. After hearing the concept for this book, Sean recommended me to Judy Bass at Industrial Press, Inc. That article is now part of the last chapter of this book.

It has been a great honor to work closely with my editor Judy Bass at Industrial Press and compositor Patricia (Patty) Wallenburg at TypeWriting to make a collection of success stories and maintenance processes into the final version of *The Maintenance Insanity Cure* and make my dream come true. Without your guidance, this book would not have been possible.

Richard Rossow, thank you for graciously agreeing to write the Foreword for my book. I am grateful for your friendship and mentoring.

Special thanks go to Mike Peterson for all his computer knowledge and IT support to make all the files and figures meet the required specifications for all the readers to be able to enjoy and use.

My sincere thanks and appreciation go to Tom Ewing, Mark Bogle, Jeff Davis, Jerry Wilson, Ron Turner, Ron Broadwater, Mike Warner, Doug Brittain, Jeff Lee, and J. T. Pundt for sharing their insights; serving as mentors; testing, applying, and refining these processes; and being there for support as friends always are.

About the Author

Roger D. Lee grew up in Louisiana but has lived in Texas, Tennessee, and Singapore and traveled all over the world. He has 43 years of experience in the chemical, refining, metals, and contracting industries with a wide variety of clients across the world. He now lives in Hockley, Texas, to be close to his children and grandchildren. He writes to share his experience and observations of life and to better appreciate all of God's blessings. He is the president and founder of RDL Solutions, LLC., a general consulting firm dedicated to helping any-size plant or company improve efficiency, cost control, productivity, and reliability. He can be contacted at RDLSolutions@outlook.com.

CHAPTER 1

How Bad Are We?

To help you feel better about your present situation, I will share a couple observations to show "what good does *not* look like":

- A Chinese plant did not have any wind socks because it looked at its distillation columns to tell which way the wind was blowing by the direction they swayed.
- A plant in Alabama would switch from the primary pump to the spare one and run it until it failed before repairing either one. This plant also waited each morning for yesterday's lab results to see what products it had made in its batch operations the day before (it was trying to make what was ordered but had to wait for lab results).

Feel better now? Both of these facilities were stuck in the insanity rut.

A Case Study

An example of a success story occurred at a co-polyester chemical plant in Malaysia. After two years of services, the demand for its product was increasing, but the plant's reliability and work processes prevented it from meeting the new demand level. This example gives you an idea of how bad things can get if the proper start-up training and processes are not adequate to prepare your people for future growth.

We collected basic plant information to define the problems needing to be resolved. We learned that:

- Product demand was increasing plant capacity requirements to go from 60% to 95%.

- The plant design uptime was 340 days per year with only three unscheduled outages.
- During the second year of operations, the plant ran a total of 180 days with 25 unscheduled outages and produced 2.5 million kg of off-class materials versus 1.1 million kg design. The plant's best consecutive days' run was 21 days, but its typical runs were 5 to 7 days before an unplanned shutdown occurred.
- Outage durations ranged from 5 to 60 days to get the plant back into operation. The molten plastic product will solidify in the piping if corrective actions are not taken within a couple of hours of an upset.

Improvement Program Initiated

An improvement program was developed to achieve the following objectives for this site:

- Achieve 100-day runs to get back to design conditions.
- Reduce maintenance spend to improve plant profitability.
- Improve plant equipment reliability while building employee capability.
- Create and implement maintenance processes to organize and add structure to daily operations and maintenance interactions.
- Create a successful site team effort involving all employees and functional groups with decisions made at the lowest appropriate level at the optimum times.
- Identify and resolve reliability improvements opportunities.
- Standardize safe work practices including permitting and lock-out tag-out (LOTO) procedures.
- Change the site culture to improve job satisfaction and employee morale.

A two-man team completed a site assessment including interviews with all functions to build our presentation to obtain management commitment. Plant-wide communications were given to share the plan and strategy, including metrics to be tracked. An update schedule was established to monitor success during implementation.

During the first few weeks, we established a Site Leadership Team with a subteam for daily operations and maintenance interactions. New roles and responsibilities were shared for all functions and levels. New work processes that focused on planning and scheduling were rolled out. Expert resources were brought in to identify and resolve reliability problems. A major effort was focused on operator and mechanic skill enhancement. The initial Malaysia apprenticeship program training had been too general in nature for developing the required skills needed. It was a governmental program to train local farmers and fishermen to become industrial operators and mechanics. Subject-matter experts were brought in from similar operating plants in other locations to deliver on-the-job and classroom materials. Troubleshooting and decision making were critical skills due to the nature of the product being produced.

One immediate change was made to impact accountability and to provide consequences. Prior to this intervention, the operators had no duties associated with plant cleanout of pluggage once the plastic set up in the lines. The operators were given required training to allow them to serve as helpers with the cleanout crews. It was now more important to them to keep the plant running.

To share progress and sustain our results, an annual milestone plan was created and shared with all employees so they could see the impact of their actions. Our training program emphasized troubleshooting and problem-solving techniques with hands-on demonstrations, preventive task skills, and strategies for building confidence for decision making. Operational and maintenance process management and condition monitoring programs were put in place.

Results Achieved

Within the first year, technical support resolved the reliability issues with the steam boiler, electrical power supply, and extruder/cutters pluggage. We added a multiskilled mechanic to each operating shift to address emergencies and evaluate identified requests to determine appropriate course of action (faster decisions and actions taken). At the start of the new processes, the plant achieved its first run of 58 days with a scheduled outage for boiler improvements and a second run of 105 days with scheduled shutdown due to high inventory. The site's maintenance and repair (M&R) costs were reduced by 49.93% after a full year of the new processes compared with the first two years' costs. This change resulted in a savings of $1.27 million realized in annual M&R spend. Key changes that produced these results included a planning process that required a work order to be written for all requests, key positions (planner, maintenance and safety coordinators for operations, and stores attendant) were selected from site employees, day-ahead planning evolved into a weekly scheduling process, predictive reliability technologies were implemented and incorporated into operator rounds with feedback to maintenance, and OJT (on-the-job training) built employee capabilities and confidence.

The plant's "maintenance insanity" cure is shown by the site manager's quote, "Prior to R&M intervention, we were not able to even get done today what needed doing. How could we possibly have time to improve? We were very skeptical of the changes they wanted to make. Now looking back after 4 years, and seeing the real results we've achieved, we are all believers."

The changes that he is referencing included

- Maintenance spend decreased by 50%; plant reliability increased by 500%.
- The capability of site employees improved to allow timely decisions with appropriate corrective actions.
- Work processes were implemented to allow repeat 100-day runs.
- Design conditions for product quality were surpassed.
- Maintenance and operations work together with joint ownership of plant assets, and overall job satisfaction improved for all functions.

A Second Case Study

This example is for a union resin plant that had been in operations since 1949 and changed owners several times until our client bought it in 2001. We put a team together for operations and maintenance to improve run-time and reliability issues. Figure 1.1 shows the typical layout of chemical batch plant operations.

Problem definition included the following areas:

- The site maintenance organization was not fully defined.
- There was not a concentrated effort to ensure maintenance provided best value to the site (work was given out more to keep everyone busy).
- The plant did not have a formal daily planning and scheduling process.
- The majority of work consisted of break-ins and emergency work (no real definition for emergency).
- The planners were not being fully utilized to do detailed planning.
- The majority of work requests were being written by maintenance (not operations).
- Communication between operation and maintenance was limited, and there was no true partnership relationship.
- The maintenance group lacked management processes for its contract support resources.

FIGURE 1.1 Batch chemical plant that broke its insanity cycle

- For problem solving, there was limited interface between maintenance and engineering.

Improvement Program Initiated

The objectives for this improvement process were to:

- Reduce maintenance spend.
- Improve plant reliability.
- Build site employee capability.
- Implement maintenance processes to organize and add structure to daily operations and maintenance interactions.
- Create a successful site team effort involving all employees and functional groups with decisions made at the lowest appropriate level.
- Identify and resolve reliability improvement opportunities.
- Standardize safe work practices including permits and LOTO and improve job satisfaction and employee morale.

An implementation plan was developed by the team to achieve examples like the following items:

- The team defined roles and responsibilities to ensure that maintenance became a site issue.
- Planning and scheduling processes required operations to be more involved with work order generation.
- The interface between operations and maintenance added communication for setting priorities, safety, equipment preparation, and needs to keep all parties informed.
- The integration of engineering and maintenance resource scheduling reduced the number of contractors for the plant.
- Day-ahead planning evolved into a weekly scheduling process including a backlog of ready-to-schedule jobs.
- Predictive reliability technologies were implemented and incorporated to provide proactive feedback on equipment status.
- Vibration monitoring was performed by maintenance.
- A site-certified welding program was implemented to increase mechanical capabilities.
- A vendor alliance was created for all storeroom purchases.

Results Achieved

Once a true partnership was created, the results demonstrated the value added by the plant's efforts:

- The site transitioned from areas to central maintenance with site priorities set for overall assignment of resources.
- The plant implemented a planning and scheduling process with 200% improvement in the percentage of planned work completed as scheduled the first year.
- The plant implemented a crew team process to drive improvement projects at the mechanic level with documented savings.
- Engineering project resource needs were integrated into the maintenance scheduling process.

- The coordination of technical support resources to eliminate problem areas through RCFA removed defects that improved run-times.
- The use of vendor alliances reduced spare parts costs.
- The site incorporated the use of value-adding reliability technologies such as vibration monitoring, lubrication, and digital reproduction.
- A site performance management program to drive site-wide cost savings resulted in $1.92 million documented project savings for the first year.
- The site reduced M&R as a percentage of the asset replacement value (% ARV) from 4.98% to 3.63% in the first year with plans to continue improvements.

The new planning and scheduling processes broke the reactive work cycle by enforcing execution of the scheduled jobs. Some of the tools that were used are provided at www.maintenanceinsanity.com for you to try as part of your improvement process. Figure 1.2 declares that schedule breaking is prohibited and gives the only acceptable reasons to ever break your schedule. Prior to breaking the schedule, be sure to try and mitigate the situation with alternative options first.

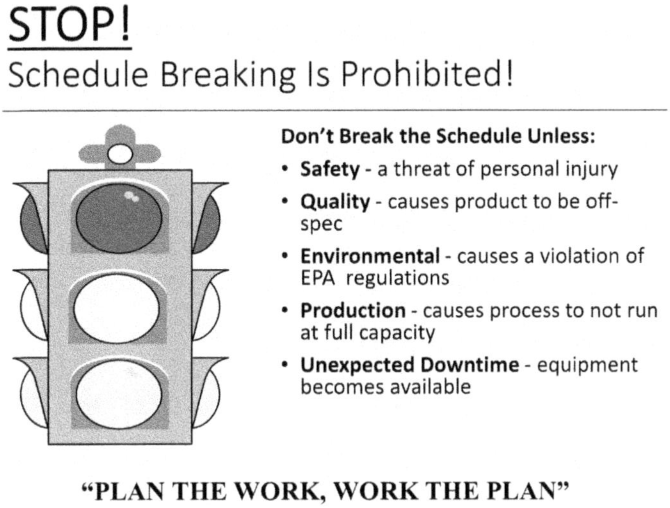

FIGURE 1.2 Enforce schedule compliance—another insanity cure

The maintenance manager stated: "The big success resulted from the reorganization with new roles to enhance communication. Operations now believes that they need maintenance to succeed or we all fail. We now fully utilize our resources before going outside for help." His conclusion was that the yearlong effort was worth the reward:

- Maintenance spend decreased by 18% ($7.39 million versus $6.06 million).
- There was a 27% improvement in M&R as % ARV.
- Maintenance FTE (full-time equivalent) was reduced from 57.3 to 37.4 with up to 6.3 FTE used for capital. Operating asset utilization increased by 16.6%.
- Work processes were implemented to increase schedule compliance (26% to 85%) and reduce E-jobs (emergency jobs), going from more than 100 to less than 10 per week.
- Pump run-time between failures increased from 13 to 26 months.
- Maintenance and operations work together with joint ownership of plant assets.

No matter what your present status may be, there is always room for improvement. Take the needed actions to address all areas of concern. To change, start by being discontented with where you are now and quit doing the same thing over and over.

Next, we will give more food for thought to see if you recognize the need for change in the insanity situations shared and see how it might relate to your own situation.

CHAPTER 2

Evaluate the Need for Intervention

Still not sure that you need to change? Consider the following stories and see if they sound like anything that has ever happened at your site. There are a lot of maintenance insanity symptoms in these scenarios.

First, we must be able to recognize the need for change. This example comes from a planning and scheduling implementation audit that was done by an experienced maintenance manager. He went with a couple of mechanics that were given the job to hang some boxes for operations. Evidently no planning was done since it was such an easy job. Once the mechanics found the boxes (45 minutes), they could not find anyone who knew where the boxes were to be installed. This job took over four hours to do what should have taken less than an hour. If the planner had been given the time to "add value" through some prework, he could have talked to the requestor, marked the location in the field (or on a digital picture), and supplied the boxes and information to the mechanics when the job was handed out so that they could have gone to the exact location with everything they needed to do the job. To some extent, these prethoughts and activities must be done for all execution forces. The planner or person scoping the job decides the value-adding information needed.

Your P&S (planning and scheduling) processes must make it easier and more efficient for the operators to know what jobs to get ready and for the mechanics to have what they need (materials and information) to start and finish jobs in the minimum amount of time.

If it is so easy to change, why do we not stop the insanity? It is because we are all change weary.

Why we cannot change:

- We have always been this way.
- We are different.

- We are overloaded.
- Nobody ever tells us anything.
- We are not all on the same page.
- Peer pressure is too negative for those who want to do a good job.

The following story does a good job of explaining why changes are needed for these typical (but exaggerated) planning and scheduling processes. Consider using it in a team meeting with your crews to see how different it is from their routine days. Capture the similarities and differences to develop your improvement plan. Why should they improve planning and scheduling when the typical data being tracked by this site indicates that they are completing 70% to 90% of the scheduled jobs (but they do not schedule for 100% of the available manpower), plus they are able to add 10 to 20 jobs on top of those listed on the schedule? The answer is "Why settle for what you are now getting when we could get so much more!" Our measures must show the true performance for all the resources. Just looking at a few pieces does not show your entire puzzle picture.

Here is an insanity test for you. You should not be able to read this text, but I bet you can (if you look and do not think about it):

The Pweor of the Hmuan Mnid

Aoccdrnig to rsaerceh at Cmabrigde Uinervtisy, it deosn't mttaer in waht oredr the ltteers in a wrod are, the olny iprmoetnt tihng is taht the frist and lsat ltteer be at the rghit pclae. The rset can be a toatl mses and you can sitll raed it wouthit a porbelm. Tihs is bcuseae the huamn mnid deos not raed ervey lteter by istlef, but the word as a wlohe.

Amzanig, huh?

You already have most of the answers locked inside you. Let this book help you see them clearly.

Based on a national survey, wrench time in an 8-hour day is 4.8 hours for best-in-class performers and 2.2 hours for the national average. Let me take you on a typical day in the life of mechanic Bill. We must focus on the human side as shown in Figure 2.1, because all reliability starts right here . . . when the skilled mechanic installs, rebuilds, or repairs any piece of equipment.

FIGURE 2.1 The human side—mechanic Bill

Crew Team Meeting Exercise: Average Workday—Pick Out the Needed Improvements

Bill reported to work on time and went straight up to the crew break area. There the supervisor gave out the assignments for the day. Bill received two jobs: one was to take care of a leaking valve on the southwest corner of the mezzanine floor, and the other was to check on a reported leaking flange on the demineralizer. The supervisor did not think they would take all day and told Bill to come back for something else to do when the jobs were finished.

This first job involved leaking stuff. Sounded pretty messy, so Bill walked to his locker to put on his older boots. Aaron was at his locker, and the two chatted for a moment while they got ready. The first thing Bill did was swing by the jobs. This was always a good idea in case a job needed special tools, or maybe the job would not require him to lug his whole toolbox to the jobsite. As he went by the first job, he easily found the deficiency tag matching the tag number on his work order. Bill had the work permit, and there were LOTO cards everywhere, so he knew it was safe to work. The valve was at chest level, so there would be no need for scaffolding or a lift truck. The valve was a 4-inch high-pressure globe valve. Bill decided to look over the other job and then get a valve rebuild kit.

At the demineralizer, the area was also cleared, and Bill had the right work permit. But Bill was uneasy. The deficiency tag was hung near a pipe flange, but Bill wondered if the line was an acid line or just a water line. In either case, Bill knew the operators would have drained the line, but it would not hurt to put on some acid-resistant gear just in case there were drops on anything.

Bill headed to the storeroom for a valve rebuild kit and to the toolroom for some acid gear. There was a line at the storeroom, so Bill changed direction and went toward the toolroom first. On the way, Bill had an idea. He knew Aaron was an experienced mechanic and had worked on the demineralizer many times. Maybe he would know if the flange was on an acid or water line. After asking around, Bill caught up with Aaron at the pump shop. After a few minutes discussing with Aaron, the two men walked over to the demineralizer. Aaron was confident that the line was only for water, and so Bill decided to skip the acid gear. It was now break time, so Aaron and Bill headed for the break room.

After break, Bill got in line at the storeroom. The storeroom happened to have a rebuild kit for the 4-inch valve. Bill took the valve kit and his toolbox up to the mezzanine floor and got to work. This was an interesting type of valve. Bill was hoping that it could be rebuilt in place. After unbolting several screws on the top of the valve, Bill was able to remove the internals but found bad news. Although Bill had the right kit to replace the valve internals, it was obvious that the valve body was shot. The whole valve would have to be replaced. The only problem was that Bill was not a certified welder and this high-pressure valve had welded connections. Bill went straight to his supervisor and explained the situation. The supervisor wanted to complete this job today and called the crew's certified welder on the radio. The welder could come over in about an hour and start the valve job. The supervisor asked Bill to return the valve kit to the storeroom and check out a replacement valve for the welder. Bill waited again at the storeroom to make the exchange, then took the new valve to where the welder was and explained how far he had gotten along. Then Bill took his toolbox over to the demineralizer to be ready to go after lunch.

After lunch, Bill took the flanged connection apart at the demineralizer. In order to obtain access to the leaking flange, he had to disassemble two other connections as well. All three flanges looked like they had Teflon gaskets, so Bill went to the toolroom for material to cut gaskets. Since he was waiting in line at the toolroom, it was a good time to call the dentist to make an appointment for next month. With the gasket material in hand, Bill went to his workbench and cut three gaskets using one of the old gaskets as a template. Bill realized that with these gaskets, he could finish up this job in no time. He wondered what the next job would be if he went back to his supervisor. It would probably be cleaning under the auxiliary boiler. He hated that job. Why couldn't he be given a pump job or something important? Well, there was no sense worrying about it. Bill gathered up his gaskets and started toward the job. On the way, he passed Gino, who was cutting out some gaskets at his workbench. After stopping to compare notes for a few minutes, they both noticed it was almost break time, so they decided just to stay in the shop and talk.

After break Bill started reassembling the flanges. Most of the bolts looked in good shape, but a couple looked a little ragged. Bill thought that he had a good handle on completing his assignment, so it would probably be a wise use of time to go to the toolroom and replace those bolts. The toolroom had an open crib for bolts, so he did not have to waste any time in line acquiring new bolts. Soon Bill finished the job, and he wiped down and cleaned up the area. He then reported to his supervisor so the work permit could be signed off and taken to the control room. By then, there was about an hour and a half left in the workday. It was customary that the crew could use the last 20 or 30 minutes of the day filling out time sheets and showering. Therefore, instead of starting a new job, the supervisor decided to have Bill go assist Jan, who was finishing up a job on a control valve. Bill helped Jan complete her job. Then he filled out a time sheet and headed to his car at the end of the day.

On the way out to his car, Bill thought about how you had to keep busy all day long just to finish one or two jobs. It just seemed that something was not right.

Did you see anything that could have been improved? Could these improvements apply to your work? Bill and his crew could have used several of the tools shared at www.maintenanceinsanity.com to have made their day a lot more productive.

So what is wrong with this picture and poor old Bill? To understand why people do not change, we will review some common characteristics that we all have:

- The number one fear is rejection.
- The number one need is acceptance.
- To manage people effectively, you must protect or enhance their self-esteem.
- Everyone wants to know "What's in it for me?"
- People hear and incorporate only what they *understand.*

Combine the time lost on these scheduled events with all your add-on and emergency work, and it is easy to see that Bill and his team are stuck in the insanity loop. How do we ever keep up? Can we improve? Now review this story one paragraph at a time and identify actions that could have been taken by the assigned P&S role (maintenance coordinator, productivity coordinator, planner, scheduler, team manager, material coordinator, store's personnel, and mechanic) to improve this typical day. This simple story shows the justification and potential benefits for improving our P&S processes.

Final Exam

Need more to prove to yourself that a change is needed? Answer these questions. How often . . .

- Do you have to wait for things while trying to do a job?
- Are you pulled off a job before you have a chance to finish it? Why?
- Do you have to wait when starting a job for operators to get the equipment ready?
- Are the permits not ready to start the job?
- Are you assigned a job that did not need to be done?
- Do you feel like your knowledge and opinions count?
- Do you take longer to finish a job than it really required?
- Are the needed parts not available when you are doing a job?
- Do you find out what you are going to work on the moment before you get your tools?
- Do you give feedback to correct a BOM (bill of material) or job plan when you learn new information?
- Are true priorities followed instead of "do what I want when I want it"?
- Does operations really know who is going to be in its area and to work on what equipment?
- Are support resources coordinated to prevent delays in doing your work?
- Do you feel like you have everything you need to do the job before you start?
- Do you know who is charging to your cost codes?
- Is all the requested work funneled through an operations representative to eliminate duplications and to verify the need and priority for each job?
- Do you start a job and then have to quit because an MOC is required to proceed?
- Do you fully understand the scope of the work you are being asked to do?
- Do you properly handle excess materials left over from a job?
- Do you get in-the-field support to solve a problem or to make an improvement?
- Do you do jobs that you think should be capital?
- Are the needed materials staged for you to pick up on your way to the job?
- Are emergencies not really emergencies?

- Are there too many emergencies to get done?
- Do jobs get put off because there aren't enough people?
- Do you ask for schedule breakers (do you have something you want to add)?
- Will operations not let you work on the equipment once you arrive?
- Are scheduled jobs bumped because "We got to have this one now"?
- Are you shorthanded for the work that needs to be done?
- Does the computer system not work properly or not have all the information you need?
- Do you "walk past it, don't see it, and don't write a work order for it"?
- Are outside resources double-booked?
- Do you schedule work that you don't intend to do?
- Are people rotated through assignments so often that no one knows his or her job?
- Does a mechanic show up to do the wrong job or not show up at all?
- Are you asked to do something before a notification is even written?

We all have room to improve. We just need to get back to basics and develop the true teamwork required to be successful. First decide that you want to survive and to be a success. Then identify and make the necessary changes:

- To adapt to economic conditions and be more competitive
- To keep our accounts and help our company stay in business
- To use more efficient ways and improve our skills
- Because what we are doing is no longer good enough

Now we will create a path-forward option to address the obstacles we just discussed and move toward our desired vision through a step-by-step Maintenance Journey.

CHAPTER 3

The Maintenance Journey

But what if I do not know where to get started to make the needed changes that I want to make for my site? You must make a change to get a change. Establish a new mindset with maintenance as an investment and not a cost. Then take the first step as described by this journey below to move away from the insanity cycle.

This Maintenance Journey describes a proven approach for changing from a reactive to proactive culture for any size facility. Processes verified by case studies shared in this book are used for implementing a successful change management strategy that ensures a smooth transition from reactive to predictive maintenance culture. A Site Maintenance Leadership Team (SMLT), a nonhierarchical team of change leaders, will drive maintenance to become a site issue with all partners involved with equal ownership. Through measures, the benefits and objectives of a proactive maintenance approach are clearly communicated throughout your entire company. The *Fish!* Philosophy (detailed later in this chapter) is one of several tools used to continually motivate and drive your people to overcome the resistance to change. Coaching and on-site support should be considered because change is not an overnight process but a continual journey, and we all need help along the way.

What does reactive maintenance look like? See if you can identify some of these elements in your present system.

Reactive Maintenance Cycle: The Backbone of Maintenance Insanity

1. Mechanics are expected to fix today what breaks today (and last night) and to address everyone's wants (squeaky wheel gets the most attention). In addition, mechanics are to check oil levels and do PMs when time is available (which it never is). In some cases, they come in early or stay late to run the waste heat boiler and operate the waste treatment facility and other utilities.

2. Mechanics come to work and are told what is on the "wants list" for that morning (this may change after the morning meeting by operations): "Go to area X and fix pump."
3. The assigned mechanic goes to the location and tries to find someone who knows what pump and what problem to be able to troubleshoot the need.
4. The mechanic returns to the shop to get tools and to look for parts or to call a vendor to order them.
5. When parts are obtained (or promised to be delivered), the mechanic returns to the area to prepare the equipment and to do the permit. This is to let operations know that he or she is starting.
6. The mechanic finally starts the job after these delays.
7. Usually prior to finishing this job, operations find a more urgent need and someone taps the mechanic on the shoulder to stop what he is doing to come fix the E-job.
8. With the cycle restarted, overtime is then required to come back and finish the initial job before going home for the day.
9. Tomorrow is the same. But if we cannot finish everything that needs to be done today, how on earth are we going to have time to plan for tomorrow?

What might it look like if we were allowed to do maintenance the way we have always wanted to?

Maintenance Our Way

1. Maintenance is established as a site issue with all parties having an active role in work identification and priority setting.
2. A core group from operations, indirect materials, and maintenance with input from HSE and technical support is assigned as the SMLT to develop the actual work processes to be followed by the site. These flowcharts are reviewed and approved by the Site Management Leadership Team and then communicated to all site personnel.
3. Critical site roles for the success for this maintenance process are
   a. Maintenance planner/scheduler for job prework
   b. Production assistant (PA) to be the single voice for operations
   c. Field safety coordinator (FSC) for third-party input and to drive the permit process
   d. Indirect materials/stores representative to obtain parts and materials

5. Daily planning and scheduling meetings are held to identify, define, and prioritize work requests and to issue tomorrow's schedule today to everyone. A one-week spreadsheet is maintained to allow moving work around to fit production needs. A weekly "look-ahead" meeting will also be held to project needs.
6. Jobs are not scheduled until all planning criteria are met.
7. The night operations crew prepares equipment and initiates permits identified on the next day's schedule.
8. The day operations crew, mechanics, and FSC finalize preparation and verify safety at the start of tool time.
9. Mechanics arrive at a job with the tools, parts, and support needed to start work. This process repeats with the next scheduled job assigned to the mechanic.
10. Supervisors assign and handle E-jobs with as little disruption of as few mechanics as possible. Operations and maintenance supervision meet prior to mechanic start time to review night activities and to identify true E-jobs that will displace today's scheduled work. Work assignments are changed prior to the crews going into the field.
11. If a mechanic is on a job and sees an additional needed repair or enhancement, he or she is empowered to do it at the end of the planned job if additional preparation is not needed. The task will be documented and feedback given to the planner for feedback to the customer. If planning is required, the job will be reported to the planner to be added to the backlog.
12. When our vision is achieved, maintenance will be in tune with the site needs and will proactively initiate corrections and enhancements at a level that there will be no need for operations to write anything other than a true emergency request.

As noted earlier, the definition of insanity is "doing the same things over and over and expecting different results."

I propose that "maintenance insanity" is what some of us are doing or have done in the past all day every day. Maintenance insanity is repeatedly doing the same maintenance tasks by the same old conventional ("We have always done it this way!") outdated methods during installation, repair, or rebuild of our equipment and then expecting by magic to have greater reliability, maximum uptime, reduced operating and maintenance costs, and better quality. It is our job to change our mindset and then share the knowledge with everyone on our sites to make converts.

Maintenance is not a cost. It is an investment. The product of maintenance is plant capacity. Through proper proactive systems using the appropriate predictive technologies, reliability is improved to ensure that equipment is available and in optimum condition to perform whenever needed. It can also reveal a "hidden" plant for extra capacity with minimum capital expenditures. Improved mechanic productivity is the human capacity produced by maintenance. The focus on reliability must be as important as the focus on safety. Doing the right things right the first time is the only acceptable option.

To be successful in changing, you must first be dissatisfied with your present condition. Even if you don't know the answer, you need to have a desire to do better. If you are happy where you are or think that you are doing a good enough job, you will not be motivated to change. Make a plan and work it. First figure out the people you need on your SMLT to drive the change and then focus on what needs to be done. Be data driven and pull the trigger. The SMLT must be the leadership for the effort and have unwavering resolve to do what must be done. Don't sugarcoat the present situation, but confront the brutal facts. Never lose faith that you will succeed. Get approval for your plan and use projected savings to justify these activities and to gain management support. Set high expectations for your organization. Do not limit your people by what you think they can do. Challenge them and let them say when you are pushing too much. Celebrate small successes to build to big ones. Share results with all employees to get their involvement. Remember that what gets rewarded is what gets done. Figure 3.1 shows the gap between where we are and where we want to be. Both leaders want to change, but they do not know how. First determine which consequences are producing the present results and change them to get what is best for the site.

The Key to Your Success: Your People

Your people are the key to your success. Focus on the human side for the most effective long-term changes to your culture. Each player must understand "why" we are changing and "what is in it for me?" All reliability starts with the mechanic at the base component level when your skilled craftsman installs, rebuilds, or repairs any piece of your equipment. Every mechanic has to know and believe that what he or she does makes a positive impact every day. It is the maintenance leadership's responsibility to make sure that "this impact" is communicated to management and to the entire site. We each need to feel appreciated for the value-adding contributions that we make. Create performance management programs that implement a fun process to capture achievements, encourage reinforcement by

FIGURE 3.1 What does it take to make it happen?

all parties, provide feedback, and develop pride in the progress being made. We have had success incorporating the *Fish!* philosophy into our performance management programs.

Fish! is a book by Stephen C. Lundin, Harry Paul, and John Christensen. The authors propose four concepts to be used to boost morale and improve results:

1. Choose your attitude.
2. Play—make work fun.
3. Be there for your customers and coworkers.
4. Make someone's day.

We have actually installed billboards and banners at the entry gates at some sites that capture this philosophy. They say:

> As you enter this place CHOOSE to make today a great day. Your colleagues, customers, team members, and you yourself will be thankful. Find ways to PLAY. We can be serious about work without being serious about ourselves. Stay focused in order to BE PRESENT when your customers and team members most need you. And should you feel your energy lapsing, try this surefire

remedy: Find someone who needs a helping hand, a word of support, or a good ear—and MAKE THEIR DAY.

To leverage the knowledge of your people, a structure must be established to link the significant contributors of a successful maintenance program into a functioning team to define and lead the standardization of your daily work processes. This will be your Site Maintenance Leadership Team. It has to have full-member representatives from operations, maintenance, and stores and support members on an as-needed basis from engineering, accounting, purchasing, contractors, etc., when appropriate for the topic being addressed. Start with the basics to gain a true understanding of your maintenance organization, crew teams, and mechanic functions by asking the following key questions:

1. What are you doing?
2. Is it getting better?
3. How do you know?
4. Can I help?

Our implementation experiences have shown that when changing maintenance, the following items are "musts":

- A work order (paper or computer) must be written for each maintenance request to properly document the site needs and workload.
- Mechanic expectations and accountability must be established and applied to set work schedules.
- Your focus must shift to reliability improvements instead of quick fixes that just keep you limping along.
- All work requests must go through a centralized planning process to utilize site resources.
- A callout process must be implemented to document actual need for overtime resources.
- Shift coverage must be minimized to get more work into the daily planning and scheduling proactive mode.

Intervention Step

The minimum intervention step required to break the reactive insanity cycle is to create a planning and scheduling function that fits your facility.

Maintenance Planning. Adding value for the mechanic through understanding and preparation of a job request prior to the start of tool time.

A planner addresses:

1. Manpower and skills required
2. Materials and parts needed
3. Support equipment and facilities to execute
4. Scope definition through job review in the field with the requestor (a picture is worth a thousand words)
5. Work instructions or stored plan
6. The task of adding planned jobs to the "Ready to Be Scheduled" hopper

Maintenance Scheduling. Assigning resources at the optimum time to allow the most efficient execution of a job request.

A scheduler addresses (with the help of the single voice for operations):

1. Making equipment available from the production schedule to take advantage of predictive technology and early troubleshooting. Gain the value in "Catch and repair" versus "Run to failure."
2. Providing required information to the site mechanics and operators the previous day to the work being scheduled.
3. Assigning the night shift to do the initial equipment preparation (block in, clean up, initiate LOTO and permits, etc.).
4. Staging materials and ensuring availability to the mechanic at the scheduled start time (if materials are not available, the job cannot be added to the day's schedule).
5. Identifying needed permits to verify safety and to drive readiness at the scheduled start time.
6. Coordinating manpower resources to staff the top-priority work for the site.

This approach uses centralized planning with decentralized execution. Change is not done to add work. We change to improve how we do our work.

FIGURE 3.2 Does it work?

Real-World Results

As Figure 3.2 shows, it may seem like a huge mountain to overcome, but through teamwork and by making one small improvement at a time, we can gain production and save the money to pay for the improvements implemented. So does it work? This process achieved the following results over a three-year period working with five companies for twelve sites:

1. M&R as % ARV improved from 4.52% to 2.80%.
2. A total of $4.8 million was taken out of the maintenance spend with improved product quality.
3. Safety performance as indicated by recordable injuries improved 69% with injuries falling from 117 to 36.
4. Savings from crew team projects were:
   First year = 24 for $321,000
   Second year = 166 for $4.1 million
   Third year = 223 for $3.98 million

We have detailed case studies for greenfield start-up, lack of plant capacity, and union and nonunion applications for U.S. and global locations. Using these maintenance and reliability standard practices, our client companies increased their asset replacement value by 41% over a 10-year period due to expansions and capital projects, while total overall maintenance spending decreased by 2%. Their M&R expenditure benchmark improved from 3.61% to 2.50% M&R as % ARV.

This group of sites improved their process reliability 19% and saved $500 million in reduced M&R expenses over a 10-year period. With globalization and acquisitions, their assets increased 41% while their total M&R spending decreased 2%. They practiced this Maintenance Journey and left the "insanity" behind using many of the tools made available for your use at www.maintenanceinsanity.com.

Summary

To make change fun and to ensure success, follow the maintenance and reliability 12-Step Program:

1. Admit that I have reactive maintenance and do not know how to stop the insanity!
2. Conceive that change is possible and that I desire sanity.
3. Be willing to ask for help.
4. Make a searching and fearless assessment of my present condition.
5. Identify and acknowledge the defects in my system.
6. Get help to develop a path forward.
7. Use data to gain management support and obtain a corporate sponsor.
8. Organize for success with roles and responsibilities for implementation.
9. Communicate the vision with all employees.
10. Establish measures and post for all to see.
11. Drive continual improvement.
12. Network to share problems and solutions.

Maintenance excellence is the missing link to being competitive in the world market today. Change occurs one step at a time, and you each can make a difference. Remember that your people hear and incorporate only what they understand. Ask for help when you need it to make the changes that you desire. Working together, we can all be successful. Build pride from your successes.

In the next chapter, we will look at specific steps to help you get started on your "cure to the maintenance insanity" at your location.

CHAPTER 4

Implementation

With the variety of company cultures represented by the maintenance industry, one size cannot fit everyone's needs when overcoming the years of insanity and bad habits. An implementation transition is desirable to establish standard maintenance and reliability practices to achieve the desired business results for you and your customers. Years of experience have been incorporated into this book for your use to impact work processes and methods, common performance measures, preventive/predictive programs, cost control, reliability implementations, budget development, and continual improvement project processes. Helpful files and tools are available for your use at www.maintenanceinsanity.com to get you started on your improvement adventure. You must coordinate the maintenance management at your sites to optimize resource utilization, improve equipment reliability, and optimize maintenance effectiveness through shared technology and practices. You can gain insight from internal and external consultants to obtain a different perspective if you are "too close to the trees to see the forest." Just getting someone from outside your groups can be eye-opening to identify and help you understand if you are in an insanity loop but may not even know it.

Using your limited implementation resources at your site, together we will drive a common strategy to take costs out of your systems, to improve labor effectiveness through better utilization of maintenance and operations personnel sharing tasks and knowledge, to eliminate low-value-adding functions, and to drive learning through networking with other facilities. As shown in Figure 4.1, sometimes you have to cut through all the red tape and make a change to get a change. It all starts with a new mindset. Believe that you can do it and you can.

FIGURE 4.1 The new mindset

The following transition process has been successful in implementing change and regaining sanity for everyone involved (self-evaluation with internal help or external professionals):

1. Visit the site to develop relationships and to assess their present status.
   a. Capture the present organization and structure to identify strengths, weaknesses, and needs. Document roles and responsibilities of all key people. Recognize small site differences that will impact the implementation path forward.
   b. Conduct an assessment as a starting reference point. Online tools and example and model assessments are included at www.maintenanceinsanity.com for your use. Collect historical data to evaluate the status of planning, shift coverage, PM/predictive programs, operator involvement with maintenance (some operators think that they are mechanics now), and other assessment categories.
   c. Do an equipment criticality assessment to establish common site work priorities and use as a future reliability reference document.
      - Assess production-critical equipment with the purpose to reduce or optimize stores inventory.

  - Prioritize business-critical equipment to most effectively allocate resources for process improvement efforts (i.e., reprioritizing or focusing improvement efforts for your company).
  - Feed business-critical assets into a reliability management model with clear prioritization for processes such as root cause failure analysis (RCFA), asset management, reliability-centered maintenance (RCM), modified maintenance strategies, etc.
  - Use the criticality assessment to re-engineer your company's maintenance processes. (Allocate resources to maintain critical assets rather than focus on low-critical, low-payback jobs.)

 d. If doing an internal evaluation, evaluate what an external consultant could offer to add value and speed of implementation.
 e. Prioritize the rollout schedule based on need versus benefit with available resources.

2. Determine the minimum resources required to meet approved implementation commitments. (We successfully used two-person teams for implementation support in many applications.)
 a. Provide experience and vision through the resources that you assign. They are to be the role models for work processes and work methods and create your credibility. They are the agents to show, tell, explain why, and change your present cultures.
 b. Define and communicate the roles of the implementation team members and the roles for the required site resources (single voice for operations and maintenance planner).
 c. Hold an initial workshop to get like-minded and to develop the following:
  - Strategy
  - Mission and vision
  - Milestones
  - Timeline with expectation

 d. Do a site-wide kickoff (town hall meeting) to initiate communications and sharing after getting approval and buy-in from the site management team.

3. The project manager and site implementation team help make culture changes in the organization, structure, work methods, and approach to problem solving:

a. Form the Site Maintenance Leadership Team with key site appointments and required organizational changes.
   - SMLT members must include a single voice for operations (production assistant), the maintenance planner, a representative from stores/purchasing, the maintenance manager, and appropriate maintenance supervision if different from the manager.
   - Develop flowcharts for site work processes and share them with all site functional groups after being approved by the Site Management Team.
   - We will facilitate this team through the maintenance manager who leads the SMLT to create a "maintenance as a site issue" culture.
   - Gain an understanding of M&R cost accounting and help with the creation of a functional location and cost center structure.
   - Drive operations involvement and needed technical support from site staff members, vendor resources, and contract services.
   - Develop and shape existing maintenance measures and visibly post them to track and drive improvements.

b. Establish a work order system (paper, database, or another computer system) including priority definitions to be shared and enforced with all work order writers. All work must have a work order. True emergency work orders can be captured after the fact.

c. Initiate day-ahead work planning with tomorrow's schedule issued to all site personnel by the end of today.
   - Establish the daily meeting routine with minimum members.
   - The planner is always working on tomorrow's jobs.
   - Maintenance supervisors and mechanics focus on today's work.
   - The store representative verifies material availability for planned work.
   - The production assistant verifies equipment availability and preparation for the start of tool time for scheduled jobs.
   - The maintenance supervisor verifies the availability of workers with the required skills to perform the work.
   - Tomorrow's schedule is populated from the planned work file.
   - True emergencies are handled with minimum impact to the schedule and mechanics needed to do the work.

d. Have the maintenance planner receive all site work requests for maintenance and capital items. This coordination of resources ensures optimum utilization of the site and contractor personnel. It also creates a management process for accountability of who is on the site for safety and invoice verification.

e. Share "maintenance awareness for operators" training to help establish their understanding of maintenance as a site issue (the true team effort required for equipment reliability and smooth operations). Basic maintenance knowledge enhances the operator's operational abilities and stops equipment damage through better understanding of mechanical, electrical, and instrument concepts.
f. Form cross-functional and crew teams to get their involvement through improvement ideas and projects that they drive. Initiate the use of a simple problem-solving model. Capture savings from their efforts and reinforce the desired results. Make a big deal out of initial successes.
g. Identify needed training for mechanics and the opportunities for multiskilling to leverage resources. Determine the options for delivery of this training.

4. Coordinate the interactions of all support personnel (indirect materials, reliability technologies, accounting, contractors, etc.). Focus on upgrading the existing PM/predictive programs through both the use of evidence-based inspections and the use of criticality assessment data and asset management index workshops to determine the optimum level of monitoring required. Evaluate third-party services and vendor agreements to ensure value for services received.
5. Document the existing spare part and material inventory and then flowchart the process being used to obtain these items. Establish a spare part setup process with justification for each new item. Leverage the use of other location surplus where it is value adding.
6. Establish networking among all the other location maintenance managers and the implementation team with monthly conference call meetings, annual face-to-face celebrate and planning functions (best practices and future milestones), and ongoing e-mail problem and solution sharing.
7. Implement a technical support project process to address identified needs. Offer pilot programs for vibration, lubrication, thermography, mechanical integrity, automation, instrumentation calibration practices, and other technologies. Possible test equipment that a site may have or want to buy to support their reliability efforts include:

a. Instrument calibrators. Checking pressures, $3,000; and input-output of electronic signals, $3,000 to $5,000
b. Electrical equipment. Megger for checking motors, $500; clamp-on ammeter, $100; and digital multimeter, $200
c. Predictive tools. Infrared temperature gun, $300; velocity/acceleration monitoring device, $1,400; and thickness measurement device, $3,000
d. Various equipment alignment devices

8. A helpful rule of thumb for estimating site maintenance staffing levels is 1 mechanic per every $5 million in asset replacement value. We have also used a 50-50 split of labor and material to take present M&R spend and show the number of mechanics being supported. Take the total spend and multiply by 0.5 to get labor cost and then divide by $60,000 (annual cost per mechanic) to get the full-time equivalents supported. Another guideline that applies to capital spends is 1.5% of ARV for business support, 0.75% of ARV for site infrastructure, and 0.3 to 0.5% of ARV for capital repairs and capital spares.
9. Monthly M&R reports are a good way for the people in maintenance to communicate to their management and to the rest of the plant personnel about their activities and achievements to add value. The reports also make a very good networking tool. A suggested format would include
   a. Safety. Give the "score" (the number of days that maintenance has gone without an injury). Share new activities and their success or failure. Include "right and wrong" digital pictures for any near miss or injury.
   b. Unplanned maintenance activities. Briefly discuss failures and solutions to share learnings and to get input from others on unresolved problems.
   c. Predictive/preventive maintenance. Share the status of PM activities (percentage complete of planned PM tasks) and resolution of action items from inspections. Discuss new predictive applications and needs.
   d. Work order planning and scheduling. Give the status (percentage of work completed as scheduled, etc.) and modifications from learnings gained. Include permitting practices and other enhancements that improve the work flow. Also address causes for delays.
   e. Cost reductions and improvements. Highlight completed crew team and other improvement projects and relationship development successes.
   f. Monthly and YTD performance. Summarize measure results and highlight new best-ever records. Give brief cost explanations of overruns and underruns and activities with accounting.

g. M&R milestone status. Give the monthly score (percentage complete) for your site plan and discuss highlights.
h. Miscellaneous items. Relate items of interest to keep others informed of personnel changes, customer feedback, programs, etc.

Establish Planning and Scheduling Priorities for Maintenance

The best way to disrupt the repeat insanity cycle is to set true priorities and stick to them when performing all requested work:

Step 1: Include customer input. This is critical customer input that is required on all notifications.

- **Desired start.** This is the date and time that the equipment will be available and ready for maintenance.
- **Required end.** This is the date and time that the equipment maintenance or job must be completed.

Specify the consequences of not meeting this end date. Planning is required to ensure that resources and materials can be available to meet the specified deadline.

Step 2: Select a priority. The maintenance coordinator/production assistant in conjunction with maintenance will select a priority from the choices provided based on the input from Step 1.

Priority 1: Emergency.* An emergency work notification form must be completed and handled according to the defined process. A management review will be conducted for each form to ensure accountability.

1. **Imminent risk of safety incident that cannot be mitigated by other means.** Mitigation might involve the use of caution tape, signs, or barricades to prevent access to a safety hazard. It also might involve moving a portable eyewash station to an area where the permanent eyewash station is damaged.
2. **Environmental impact that cannot be mitigated by other means.** For some applications, a bucket or pan placed under a drip will mitigate an environmental problem until a repair can be planned and executed.

3. **Shutdown of operating equipment that must run today to meet the business plan or to meet regulatory requirements.** This does not apply to process equipment that can be safely bypassed and operated until a planned repair can be made.

* An emergency is when a real and immediate threat exists to life, health, or property. Substantial production losses will occur if immediate action is not taken. By definition, emergency work is performed without formal planning. Work is to start immediately and to be carried through until the malfunction is corrected. Overtime is authorized. For all other priorities, overtime is to be scheduled by authorization only.

Priority A: One to three days (moderate safety risk or potential production loss). A malfunction exists that will generate a Priority E (emergency) if not addressed within 72 hours of notification.

Conditions include

- Priority A (urgent/next day) notifications identified early today and able to be properly planned will supersede Priorities B and C work to be done on the next day's schedule.
- If the Priority A request is submitted after 2 p.m. and must be completed the following day, this request must then be converted and handled as a Priority E including the required additional documentation and review.

Priority B: One week (redundant equipment and some predictive and PM work). This is a request that can be handled in the regular course of maintenance performance and services as populated on the schedule, but it has been identified as needing completion within a set time frame due to miscellaneous reasons.

Priority C: Two to three weeks (low-risk safety-related work, most predictive and PM work and improvements). This is a request that can be scheduled for the regular course of maintenance performance; services and preventive maintenance tasks that populate the schedule as resources are available: calibrations, ISO, MI, SIS, etc. Maintenance specifies the timing and need for overtime.

Priority D: Fill-in (one week to three months) and capital work. Fill-in jobs require little or no preparation by operations or maintenance to perform. Only a safe work permit is typically required. These jobs can normally be started and stopped with little loss in effectiveness. The planner will estimate the labor-hours required to complete. Capital jobs have to be scoped out to allow the maintenance crew to be pulled off as required.

Priority S: Shutdown. This work can only be performed during a plant turnaround or equipment outage. Plant turnaround refers to the time during the year when processes are shut down to allow for preventive maintenance, modifications to installed equipment, and new equipment or building installations.

Safety and regulatory note: Work that has to be completed to address a law or regulatory or safety concern needs to be worked into the schedule ASAP but not on an emergency basis. True safety and regulatory concerns must have adequate detail provided to be evaluated for immediate threat of life, health, and property.

Step 3: Set and communicate priorities. The production assistant/maintenance coordinator will review the work notification listing on a weekday basis for setting priorities with maintenance and will communicate these priorities to the maintenance planner and scheduler.

Step 4: Review backlogs to ensure priorities. The maintenance planner, scheduler, and supervisors will review the work order list on a weekday basis and review the backlog on a (minimum) weekly basis to ensure priorities are managed. The work order backlog review is to be part of the week-ahead scheduling meeting for customer input and concerns. Backlog searches can be made for different date ranges.

Check Step to Verify Progress

Once you have everyone participating, the last step to continual improvement is to periodically check and see if you have really changed.

For the vision of the planning and scheduling processes to have optimum positive impact, the following changes must occur:

1. A true partnership among operations, maintenance, and stores exists with improved communication that results in more efficient work being performed.
2. The maintenance coordinator (MC) filters all notifications and feeds more complete information to all maintenance forces with as much lead time as possible. Lead operators support the MC in permit and equipment preparations especially during outages.
3. Planners minimize today's activities to only quick value-adding contributions and focus on tomorrow and beyond to make job packages for all planned jobs.
4. The maintenance team manager works in the field supporting and building mechanic capabilities and resolving execution concerns.
5. The scheduler (not the planner) focuses on coordination and communication with all resources to create and execute the daily schedule.
6. The material coordinator (relief planner) is the first point of contact for hustling parts and materials for emergency and planned jobs. Mechanics continue working on assigned jobs while the materials are being obtained and brought to them to expedite completion.
7. Backup relief is in place for all key players in the P&S processes. Relieving each other or covering two jobs does not allow progress to be made.
8. Mechanics provide feedback for continual improvement to BOMs and job plans.
9. We schedule 100% of available resources daily with fill-in jobs available as a contingency.
10. We are getting better at saying today what we are going to do tomorrow and sticking to it.
11. Maintenance knows who is coming to work in its area tomorrow and communicates it to operations for night shift preparations.
12. The focus on daily schedules based on labor-hour job estimates is resulting in more work being performed.
13. Flexing within areas is allowing requested work to be done with less overtime. Designated resources are freed up in advance to prepare for shutdown work.

In the next chapter, we will add more structure to create a true planning and scheduling process to ensure we stop doing the same things over and over by fixing small issues before they become the big problems that keep us trapped in the maintenance insanity cycle.

CHAPTER 5

Planning and Scheduling Guiding Principles

When everyone is in charge, the truth is that no one is really in charge. Insanity thrives without proper communication and cooperation. It is critical that the scheduling process be jointly owned and managed by both the production and maintenance departments with input from all parties that want work performed in the operating units. No single group has all the knowledge required to create the optimal schedule. Production must provide information on optimal timing to meet production requirements, and maintenance must provide information on optimal timing to meet worker (site and contractor) and material requirements. In order to schedule effectively, you must have enough lead time to arrange for all the required materials, preparations, and resources to be provided prior to the start of the job.

A weekly scheduling meeting is a method of achieving the required advance notice for requested work. This is to be a one-hour meeting held Wednesday afternoon. The planner's backlog of planned jobs in the RSCH (ready to schedule) status and other backlog work requests are to be reviewed in this meeting along with any new activities from the engineering capital project timeline spreadsheet for next week. All interested parties with desired work for the next week are invited to attend to give their input. The result of this meeting will be the weekly schedule for the coming week, which will be published to the site on Thursday. The weekly schedule should contain enough work to allocate 80% of the available resources for the following week. The remaining work will be filled in on the daily schedule to handle Priority 2 work that can be expected to occur during the week.

These two communication tools will keep all parties informed of upcoming work. The weekly schedule shows all major jobs that are expected to be executed in the upcoming week, and the daily schedules that are given to operations in the afternoon prior to the next day's work show the exact jobs that need to be prepared for execution in the morning.

All work identified and requested should be evaluated to answer the following questions:

1. Is the work necessary? (Is this a duplicate request? Does it add value? Is there another option?)
2. Can the original scope be reduced to achieve the same or better results? Can we combine jobs to prevent multiple work orders or contractor purchase orders (POs)?
3. What can be done to prevent this job from occurring in the future? (Is it possible to make changes in the design of the system or work method to prevent future failures? How can the defect be eliminated? This step has the most potential for real savings but is often ignored in the planning process.)
4. What is the true plant priority for this work? (Compare all jobs relative to site production demands and readiness to be performed.)
5. How much planning is needed? (Does the mechanic know how to do this job? How much value can be added to the efficiency of his or her wrench time? Does a job plan already exist?)
6. Are long delivery purchases involved?

The planning process will address parts to order and stage in advance, required special skills and equipment needed, extra resources or skills required, safety hazards that exist, permits required, and the most efficient work method. Planning provides all the known necessary materials, information, and resources to do the job correctly. Each job must be evaluated to determine the amount of planning that will add value to that specific job. No more or no less effort should be applied to plan that job than is justified by the end result of making the mechanics work more efficient and eliminating delays.

The daily planning and scheduling strategy has three goals: increase plant reliability, improve mechanic tool time productivity, and reduce maintenance cost. Efficiencies gained in planning and scheduling do not go to the bottom line unless the savings are actually removed from the maintenance cost structure by reducing external resources, performing more work with existing resources without working overtime, redeploying internal resources or work tasks, or reducing material purchases or overall cost. A true team effort involving operations, maintenance, engineering, stores/purchasing, and safety is required to make maintenance a "site issue." Your Site Maintenance Leadership Team must develop flowcharts for the maintenance work processes that will provide the structure to make our strategy a success. These will be communicated to all site employees as soon as the flowcharts are completed and approved by site management.

The craft worker's role for obtaining materials is evaluated for each job. Craft workers are expected to get their own materials for emergency needs if it is the most efficient option. If they have value-adding work that can be done while the materials are being obtained, contact the material coordinator to get them for you. Materials should be staged or kitted for planned jobs.

The following simple forms can be very beneficial in breaking the insanity loop by providing the right information to the mechanics to allow proper preparation before the work is started so that they can do what needs to be done instead of what they have always done. Insanity practices cause Band-Aids to be applied, but planning and scheduling leads to permanent solutions and process improvements. Several P&S flowcharts and useful tools are provided at www.maintenanceinsanity.com to give you ideas to consider.

P&S Audit and Feedback Form

Job Package Audit Date: ______________

1. WO # and job description: ______________________________

2. Score the information you were given: 0 = None to 5 = Just what I needed
   SCORE—did you get any information "Y or N" and how good was it "1 – 5"

   - Was there a printout of the work order long text or other scope description? (Y/N) ______
   - Did you understand what to do to solve the problem? (Y/N) ______
   - Did you understand when this job is to be completed? (Y/N) ______
   - Did you know the exact location and when the job is to start? (Y/N) ______
   - Were materials provided or staged when the job was assigned? (Y/N) ______
   - Was safety and permit information provided for the specific job? (Y/N) ______
   - Was the name of the LOTO/area contact provided? (Y/N) ______
   - Were needed support resources available at the start? (Y/N) ______

- Were pictures, sketches, or drawings provided? (Y/N/NA) _______
- If personal lifting was required, was the weight specified? (Y/N/NA) _______
- Was the test equipment calibrated and provided? (Y/N/NA) _______
- Was the equipment down and ready at the scheduled start? (Y/N) _______

3. What other questions do you have that could have been answered prior to starting this job that could have made your work more efficient or eliminated delays or problems?

- ______________________________
- ______________________________
- ______________________________
- ______________________________

4. Now that you have finished this job, what feedback can you provide to the planner that will make it easier to do next time?

- ______________________________
- ______________________________
- ______________________________
- ______________________________

Audit Performed by: ______________________________

This second form can provide a good source of information from the mechanics to help the planners make each repeat job better the next time it is performed.

Mechanic Job Feedback

WO # ________________ Date: ________________

WO description: ____

____

1. Was this job an emergency, was it a break-in, or was it scheduled? (E/B/S)
2. Was a job package provided when the work was assigned? (Y/N)
3. Were the equipment and permits ready for the start of work? (Y/N)
4. Identify causes of delays and actual time lost: ______________________

   ____
5. Did the information shared from the planner, team manager, or scheduler help you do the job more efficiently? (Y/N)
6. What else did you need to know? ______________________________

   ____
7. What parts were needed that were not provided? ______________________

   ____
8. Did stores have the materials you needed to do the job? (Y/N)
9. Who helped you obtain these needed parts? ______________________
10. What corrections need to be made in the BOM or stored plan?

    ____

    ____
11. Was the work order maintenance activity on the repair type (REP, PRV, PRD) marked properly? (Y/N)
12. Did you mark the breakdown flag when an equipment component was replaced or modified to get the equipment to perform properly? (Y/N/NA)
13. If yes, identify the failed component that was the primary cause of the breakdown: ______________________________. Complete the technical report.

14. Document solutions to prevent future failures:________________________________

____

____

15. Did you provide job status to your team manager by 2 p.m. if the work required overtime to complete or had to be carried over? (Y/N/NA)

16. Did you enter your time for this work order? (Y/N)

General Feedback: Please let us know of any problems or suggested improvements that are needed to make the planning and scheduling process work better.

____

____

Our experience in moving from reactive to proactive with effective planning, scheduling, and execution has shown increased productivity by about 32% compared with the reactive mode of operation (typical breakdown is 2% tools/travel, 7% waiting for permits and operations, 4% obtaining job lineup, 3% obtaining materials, 3% obtaining drawings/procedures, 3% inefficiency, and 10% unassigned).

Work requests are evaluated by the criteria shown in Figure 5.1 to determine the level of planning required for each job. All jobs are not equal. Apply the appropriate time and attention to get the most from your limited resources by doing the right level of planning for each job.

The following list contains the essential elements that will maximize the success for any planning and scheduling process:

1. Notifications are to be written as soon as the need is identified to allow as much lead time as possible to plan the work. The lead time concept must be understood and implemented to change from our reactive mind-set. Challenge the comfort zone by setting a level of risk that we are willing to accept.
2. Work notifications are to be separated into types (emergency, routine, shutdown).
3. Operate with centralized scheduling for defined units of business (site, division, etc.).

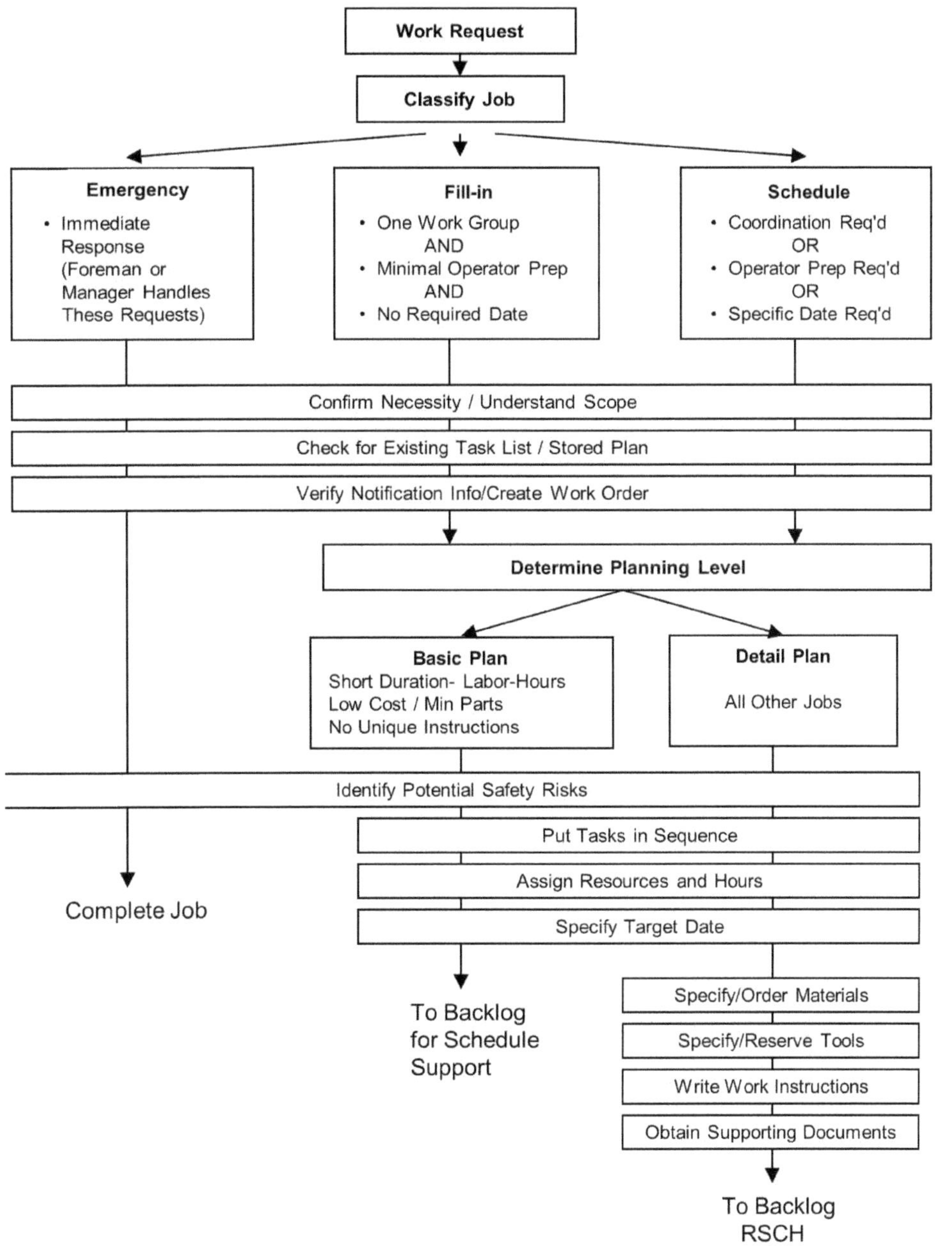

FIGURE 5.1 Planning decision tree: selecting the level of planning

4. Planners have to focus on "tomorrow" and be isolated from "today" activities.
5. Dedicated shutdown planners are required for defined outage sizes (set a labor-hour or cost limit)
6. An operational function (maintenance coordinator, production assistant, etc.) is required to serve as a funnel and filter for work requests. A productivity coordinator role performs the same function for the execution forces.
7. Limit the number of work notification creators to the critical few.
8. The P&S process has to be an equal partnership described by the three-legged stool model.
   a. Operations input for when the equipment is available to do the work
   b. Supervisor/scheduler input for who is available to do the work
   c. Planner/stores input for what materials and resources are available to do the work
9. Roles and responsibilities for required functions are to be defined and enforced with known accountability.
10. A set number of mechanics can be reserved to handle emergency work. This could be achieved in a variety of forms (ops mechanics, emergency crew, etc.). This would remove the emergency job burden and create a manageable planned work backlog.
11. The planning and scheduling functions are to be separated (even if they have to be done by the same person in small sites).
12. The use of standing orders for planned work must be minimized or eliminated. For nonplanned work, this is an acceptable method.
13. To truly understand our level of performance and to identify improvement opportunities, we should schedule for 100% of available resources with a buffer portion of fill-in work to allow for flexibility.
14. Implement consequences for requesting emergency work, especially if you exceed the dedicated resources assigned to handle E-jobs. Have a special form required with an accountability process to follow up the next morning to verify that this was an actual emergency according to a set definition.
15. Workers and other resources must be shared across boundaries (crews, groups, departments, divisions, etc.).
16. Establish a site prioritization system with specified criteria and basis (criticality of equipment, production demands, HSES issues, etc.).
17. Evaluate work order historical data to flag and follow up on abuses of "not urgent and not important" requests.

18. Create a toolbox of IT systems available for P&S implementation (SAP, Access, Excel, etc.) and provide training support as areas implement the best-fit tool for their needs.
19. Define "work to be planned" to demonstrate the value that will be added by this activity. By default, recognize the "no plan needed work."
20. Define the standard performance measures (select a few) to be used to drive P&S.
21. Enforce a feedback loop to continually improve job plans and the overall P&S process.
22. Create the guidelines for a standard backlog management strategy and implement.
23. Implement and enforce a schedule break process with follow-up accountability to justify actions. Give this the same emphasis as safety and cost control efforts.
24. Create a standard approach for P&S that can be modified slightly to fit every application. It is no longer acceptable for each maintenance group to do something totally different from everyone else.
25. Daily schedules must include all work requiring coordination that is being done by groups charging to maintenance. This will provide a complete list of next-day jobs to operations for preparations during the night shift.
26. It is up to operations to identify a needed job based on data that justifies the work and to provide as much detail as possible to define the problem. It is up to maintenance to decide how to make the repair or improvement.
27. Some P&S models can be used to optimize existing resources (identify roles needed), to visualize potential savings to our customers, and to help sell and justify implementation of our process.
28. P&S tools must link to your site computer systems (SAP, etc.) and be standardized enough to be supported. Data purge processes must exist to prevent the slowdown of computer response.

Be sure to set appropriate boundaries for P&S improvement processes before charging your improvement teams to make the desired enhancements. Consider these as your starting point:

1. The improvement teams cannot add more people to the organization but must make their proposed changes within existing head-count limits.

2. The planning and scheduling process guidelines given in the text above are to serve as the starting point for the development of department processes. It is expected that this will cover 80% or more of the department needs.
3. The P&S Improvement Teams will customize these processes to fit area-specific needs and then make proposals to their Steering Team for approvals of modifications prior to rolling out to the entire department. They are to:
   a. Specify who will handle each of the required functions.
   b. Determine if more than one role is to be done by the same person.
   c. Recommend needed changes in present assignments and duties.
   d. Propose the method for handling emergency work without P&S.
   e. Provide a process for the interface with all needed support resources.
   f. Develop workflow processes for your unique area requirements.
   g. Establish daily communication meetings to share needed information.
4. To get the basics in place, an initial spreadsheet must be available for daily emphasis when working with the weekly schedule.
5. The P&S champion will work with each Manufacturing Division Team to establish the appropriate makeup of a Steering Team to approve P&S Improvement Team recommendations and to support cultural change issues.
6. A vision document should be provided for this P&S initiative.
7. If an OP Team (mechanics doing direct operational support) arrangement is appropriate, employees are not to be transferred out of the maintenance organization.
8. Processes dealing with shared resources (shops, scaffolding, insulation, etc.) should be consistent from one area to the next.

Part of the cure to the maintenance insanity is setting boundaries and following rules and established structure for P&S. The next chapter focuses on the basics that must exist for understanding the need for change and then achieving it through good communication and teamwork.

CHAPTER 6

Back to Basics

So that I do not insult your intelligence, I will use an example with monkeys that does the best job of describing why the insanity seems to continue forever. Let this be your wake-up call to see if your group needs a change.

Monkeys in a Cage: Is a Culture Change Required?

To keep from hurting anyone's feelings or stepping on too many toes, I can honestly say that this example is not about any maintenance organization (but it might contain some learnings for your present situation).

Start with a cage containing five monkeys. Inside the cage, hang a banana on a string and place a set of stairs under it. Before long, a monkey will go to the stairs and start to climb toward the banana. As soon as it touches the stairs, spray all the other monkeys with cold water.

After a while, another monkey makes an attempt with the same results—all the other monkeys are sprayed with cold water. Pretty soon, when another monkey tries to climb the stairs, the other monkeys will prevent it. Now, put away the cold water. Remove one monkey from the cage and replace it with a new one.

The new monkey sees the banana and wants to climb the stairs. To its surprise and horror, all the other monkeys attack it. After another attempt and attack, it knows that if it tries to climb the stairs, it will be assaulted. Next, remove another of the original five monkeys and replace it with a new one.

The newcomer goes to the stairs and is attacked. The previous newcomer takes part in the punishment with enthusiasm! Likewise, replace a third original monkey with a new one, then a fourth, and then the fifth. Every time the newest monkey takes to the stairs, it is attacked. The monkeys that are now beating him have no idea why they were not permitted to climb the stairs or why they are participating in the beating of the newest monkey.

After replacing all the original monkeys, none of the remaining monkeys have ever been sprayed with cold water. Nevertheless, no monkey ever again approaches the stairs to try for the banana.

Why not? Because as far as the monkeys know, that is the way it's always been done around here!!!!

This example shows the need for explaining "why" for a proposed change and "what's in it for me." Perception that things have always been this way and will not be able to change has to be addressed in terms of "know-how" and "want to." Training, coaching, and showing are needed to address the "know-how" issue; "want to" is more of an attitude and trust issue. Learn from this lesson to resolve concerns and to gain understanding, and then we will all be willing to give it a try when we understand more about the proposed change.

Here is a maintenance example to help make the point. At one site, the union had very specific craftspeople like steam fitters who could only work on pipes with steam in them. P&S was being done to keep each craft busy instead of planning and scheduling for the most needed work and using the most efficient resources to complete the scheduled jobs. Once we studied the union agreement, the mutual decision was made per the contract to have craftspeople serve as helpers if their specific skill was not fully booked on the day's schedule. The change had never been considered because the mind-set of "This is the way we have always done it" did not allow options to be seen.

Communication and Teamwork

But what if my site is nowhere close to being ready for any of these suggested drastic changes? Then we can help create the steps to start your journey at your present stage of P&S maturity.

The key to improving productivity and maintaining cost control is communication and teamwork. This process is a commonsense approach to doing what we all know is the right thing to do. Assume that there is a dome covering your plant. Planning, scheduling, and execution must apply to everyone entering the dome to do work. Open and honest communication and coordination of resources are the two critical elements for success.

Operations as the end customer must be willing to "say what you want" and then *stick to it* to allow maintenance forces including contractors to "do what you said." Everyone understands this concept for a race team to be able to win a race. In most plants today, the maintenance and operations relationship is more of a garage versus pit crew approach. There is an overall lack of communication from both directions. We do not listen to each other, nor do we work as a unified team.

We are all resistant to change because we believe we already know the answer or because the proposed solution has been tried before and did not work. Our plant structure encourages each division or department to be different, resulting in lack of standard practices and unity. Successful change management must have owners from each organization. Client owners must be established for oversight of contractor forces for all tasks being performed by them. When contractors do operational support functions, maintenance does not feel ownership. An owner from operations or material handling must be named. Most contractors are not fully plugged into the client's planning and scheduling processes, and few contractors have true planners. There is even a lack of standardization for the key players in P&S for all participants.

Nothing changes until the shop floor does things differently. Our employees and contractors have the right to do the job right. Take care of the place where you make your living, so it will take care of you. Our goal in aligning maintenance, including contractors, operations, and stores, can be covered in the following statements: We all want to get more of the most needed work completed with available resources in the most efficient way. We want to make repairs and improvements so our equipment runs better and longer between failures. And we want to minimize negative impact on production without compromising safety.

The diagram shown in Figure 6.1 describes how a race team is structured for success. All activities are contained within the large outer circle and focus on winning the race. Sharing the right information through clear communications is critical to keep the car running to finish the race with minimum delays in the pit. The players have to listen to each other, analyze the data, and make the best decisions for the situation. They win or lose together as a team.

Your organization can make the necessary changes to create the same structure and work processes for your maintenance and operation partnership to create this "under the dome" approach. In Figure 6.2, the Site Leadership Team structure is set up to manage all work coming into the "dome" that covers the entire plant. Managing this way will take a new mind-set. You must make a change from the present insanity to get the new desired change. Maintenance services should not be looked at only as a cost but as an investment for your plant reliability! The product of maintenance is plant capacity. The focus on reliability and productivity must be as important as the focus on safety.

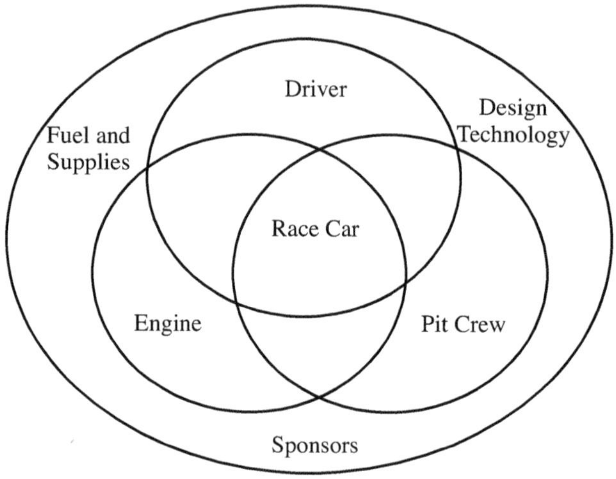

FIGURE 6.1 Race team structure for success

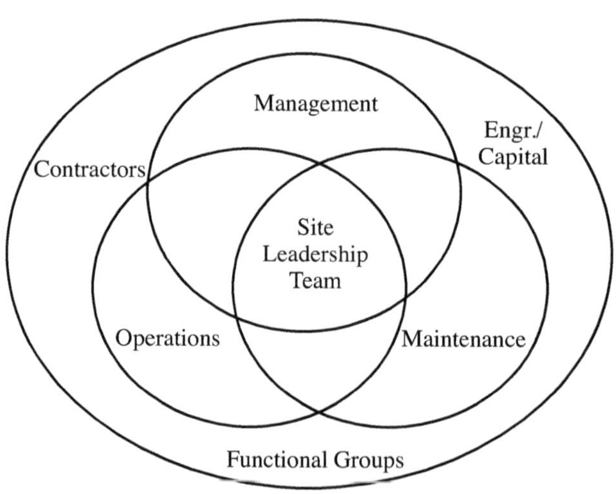

FIGURE 6.2 Site leadership team structure

All P&S prework and preparations only add value when they are communicated to and understood by the craftspeople, mechanics, and operators. Each participant must use the information and scheduled resources to make job planning count and be effective.

Leadership Attributes

Where maintenance insanity is out of control, no one stands out as the leader, and everyone just seems to go with the flow. The members of the Site Maintenance Leadership Team must stand out like a Georgia mule in the Kentucky Derby. Our people need leaders they can respect and from whom they can expect to receive the guidance needed to move to the vision being communicated.

The following observations are provided to describe leaders that earn respect and know how to guide their organizations to success. Leadership is imperative for the success of any effort. We are all leaders at different times. Step up when it is your time to make the needed changes.

1. Be a servant leader and engage your people. Make eye contact and really get to know them. Be judged by what you do and by what others can do with the help that you provide.
2. Complaining gives other people power over you. If it is easy to say, keep it to yourself.
3. Tell your people how good they are (specific examples) and tell everyone else too. Focus on the positive to change perceptions. Tell them why you are proud of them. Value your people.
4. Make people and their actions priceless in your heart. You will treat them differently, and they will react to you differently. You are the only person that you can change. This time will never come again. Get busy caring about others. Your employees have been entrusted to you. Never make them wonder if they can depend on you.
5. Know your job and your people. Publicly take blame for the bad stuff under your responsibility and always give away the credit for the good. Be willing to make decisions with integrity and to take action.
6. Leadership development is an investment in the future. It is always your job. Attitude overcomes obstacles. Be fired up. Self-confidence allows us to have confidence in others to let them grow. Underwrite risk to learn from mistakes. Own up to your own mistakes, and others will also. Employees expect you to do your best and for you to take responsibility.

7. Never do anything at the expense of others. "BS" may get you to the top, but it will not keep you there. Be a pleasure to work with. Treat others as you want to be treated. Be the boss you have always wanted to have.
8. Learn to listen so you do not miss the good stuff. Pay attention to the input, understand the meaning, evaluate importance without judging, and respond appropriately. Saying nothing may be the best response. Be aware of personal bias and stay open to new ideas. Practice patience. Listening helps to build relationships. Use your eyes to hear the complete message.
9. You cannot lead from an office. Build relationships. Excuses keep you right where you are. Leaders create the vision. When you get where you are going, where will you be? Be open to learn and see different realities. What can you do to help your people? Motivate through gratitude by showing true appreciation. Your face speaks volumes about you. Wear a smile. Learn to laugh and celebrate everything.
10. Get a checkup from the neck up and be the best person you can be. Lead by example. Let go of the past to move forward. Forgiveness is a decision only you can make. Care about people and be sure to tell them. Live in possibilities and not probabilities.
11. It is OK to say you do not know. Have a support network to find answers. You can trust others and still be in charge. Get input from all levels to find out what is really going on. Leverage knowledge to create give-and-take partnerships. Learn daily.
12. Begin with the end in mind. Leaders create the future and take us with them. See it and then do it. Find something to be passionate about. Nothing happens without communication. Optimism is a force multiplier that gives hope. Coach the players to win the game.
13. Character in leadership is like wet to water. Humility of spirit keeps you rooted with an overflowing cup. Leaders are shaped by experiences and practice. Exercise your brain to keep it alive.
14. Maintain a servant's heart to build up your people through your leadership and your relationship with them. Work together for the benefit of all. No one is better than anyone else.
15. Invite success by living in a full-court press—make your work your "calling." Be an inspiration to others. Be someone's hero. Use time to your advantage, never get too old to learn, and be rich in friends and love. Your children are your living messages to a time you will not get to see. Realize

how good life is while you are living it. Touch someone and leave a mark on the person's life. Fully utilize the helping hand on the end of your arm. If not now, when, and if not me, who?

16. Use the PUNT process to see the perspective of the other person to understand each other to negotiate the best solution for a trusting relationship.
17. Leadership is having concern for others, doing things collectively, and showing by example. Our parents are the first leaders in our life. Lead in a way that creates followers that will do what they think you would want even when you are not there. This occurs when your people are your first concern.
18. Laughter brings the body into balance. The healthiest emotion is gratitude. Laughter puts gratitude into attitude and will overpower worry and concern. The worst emotion is vindictive anger and only leads to stinking thinking. Be yourself to reduce stress. Live so that you are all used up when you die. Enjoy the life you are given and make the most of it. See the humor in everyday life. "Stressed" spelled backward is "desserts," so now you know the answer to relieving it. Life by the inch is a cinch. Over 90% of the things we worry about never happen.
19. To get stress "payoffs," go through the motions expecting the worst but preparing for the best and appreciate whatever results occur, being grateful that you could participate. Have faith knowing that in the end we have the opportunity to go to a better place.
20. Internalize little goals that are achievable that lead to the overall result you want. Act unafraid and motivated. Fake it until you make it. Carry a song in your heart and a smile on your face to accept problems as opportunities to make you feel good about yourself.

Once we have our leaders in place and they have developed their plan, be a part of the success by playing your role and helping others achieve their goals. Start identifying and removing obstacles to ensure you have all the basics in place.

One of the biggest deterrents to change is allowing everything to be called an emergency even when we all know that it is only a want. How can a leak that started as a few drops per hour and has progressively increased in volume over the last month be called an emergency when someone finally realized that it is making a mess in his or her area? *We must agree on a mutually beneficial definition for emergency maintenance and then enforce it religiously.* To achieve maintenance efficiency, we will never break today's maintenance schedule unless we have a true emergency need. A true emergency maintenance need would be one of the following (as first noted in Chapter 4):

1. An incident where there is an imminent risk to safety that cannot be mitigated by other means. Mitigation might involve the use of caution tape, signs, or barricades to prevent access to a safety hazard. It also might involve moving a portable eyewash station to an area where the permanent eyewash station is damaged.
2. An environmental impact that cannot be mitigated by other means. For some applications, a bucket or pan placed under a drip will mitigate an environmental problem until a repair can be planned and executed.
3. The shutdown of operating equipment that must run today to meet the business plan or to meet regulatory requirements. For some processes, equipment can be safely bypassed or operated until a planned repair can be made. Some equipment can be partially bypassed to relieve pressure and then continue to operate until repaired.

For emergency work, the maintenance crew will start immediately and work nonstop until the job is completed. This means that overtime is approved as required and permission is given to expedite parts and materials at premium cost to speed up the resolution of the emergency. Emergency work will bump planned work for that day. This type of work is the most expensive and least efficient means of doing maintenance. All efforts must be made to minimize the amount of emergency work.

With more attention given to preventive and predictive maintenance (especially more observation by operating personnel to recognize and act on problems before they become emergencies), we can drive the need for emergency maintenance to essentially zero (<5% world class). Less emergency maintenance means that more value-added maintenance will be executed following our P&S process!

Putting these basic components in place will be just what the doctor ordered to treat maintenance insanity. The next step will be to cultivate and nurture the true partnerships needed to support our cure.

CHAPTER 7

True Partnership

The basic step to success is having a true partnership with operations, stores, and maintenance. This is what I call the three-legged stool model as shown in Figure 7.1. If any one of the legs fails to fully function, the stool cannot hold your weight and you fall. Maintenance must be built on a trust relationship. All the parties involved must "say what they mean and mean what they say." We only get in trouble when we try to read each other's minds. Maintenance insanity flourishes when we make assumptions and stop talking to each other.

This partnership is based on the following key roles being performed by the appropriate people in your organization. Each of these positions must be filled by the best person for each role. These descriptions can be changed to fit the selected male or female.

Maintenance Coordinator/Production Assistant

The maintenance coordinator (MC) is the operations representative and spokesperson who communicates the production needs to maintenance with as much lead time as possible to allow for adequate planning and scheduling. Because the MC's opinion is trusted by all operating personnel, he sets the overall priority for all work assigned to maintenance. He collects work requests (notifications) for his entire area, filters them for duplicates, edits them to best define the needed work, and then shares all known information such as troubleshooting feedback with the planner. He uses the work notification backlog to review all new requests from the prior day.

The MC keeps the scheduler informed of production changes that may impact the schedule. He has ownership for communicating the night preparation needs required to have identified equipment ready for the permitting and LOTO process verification in the morning to prevent job delays as work gets under way in accordance with the day's schedule. He is supported by the utility (day) operators to handle heavy schedule loads,

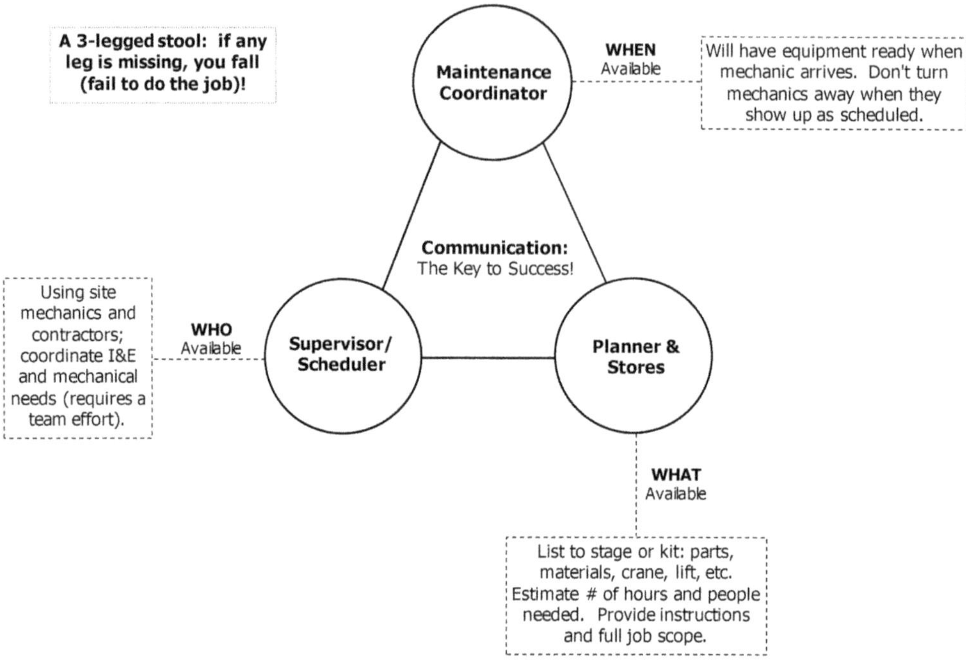

FIGURE 7.1 Three-legged stool maintenance model

shutdown work, and upset conditions. He ensures that all preparations are ready at the scheduled job start times for all resources listed on the daily schedule. He handles preparations for nonscheduled resources after he has completed those that were scheduled.

The MC is a key member in the daily communication process to answer questions and make necessary contacts to ensure tomorrow's schedule can be done and is published by the end of the day. He advises the scheduler each day on the most needed work that fully loads the available resources. He is also critical in responding to emergency needs to get the failed equipment ready for maintenance. He is the consistent voice for operations to get the needs met for all shift crews and day personnel.

Contractor/Client Productivity Coordinator

The productivity coordinator (PC) has complete responsibility for ensuring that all requirements and work activities performed by the field personnel are conducted safely and cost-effectively and with the proper quality to improve crew overall efficiency. This position works closely with the contractor supervision, contract services, and the maintenance coordinators in the field to reduce barriers that affect the productivity of craft personnel. The productivity coordinator is the number one controller of field productivity, ensuring that the weekly frozen schedule is adhered to and deviations are minimized.

The PC will give input to the development of the client weekly sequenced schedule (including all contractor work) to help the crews bundle work and stage materials. Bundling input includes demos of scaffolding to use for new builds, the grouping of hard- and easy-to-achieve jobs, jobs in the same immediate area for material staging, and jobs to best fit the use of rack trucks or buggies.

The PC helps achieve area customer buy-in to gain acceptance for the proposed contractor input to the schedule job sequence. Another responsibility is to ensure that accurate information is being shared with all client and contractor crews. This addresses maintenance and capital work as well as operations support. The productivity coordinator position requires approximately 75% of work-hours to be spent in the field interacting with the crews and the remaining 25% in the office.

Scheduler

The scheduler owns the schedule (daily and weekly) from creation through execution. Her focus is on getting all the right resources to show up at the right time to get the most work done in the most efficient way. This is not just an administrative position created to publish schedules. The scheduler juggles people and support equipment as necessary to execute the issued schedule each day and then takes today's feedback to adjust tomorrow's schedule prior to the mechanics going to the job.

The scheduler is the main interface for all execution forces coming to the area to perform work. Her schedule reflects all their preparations and permit needs at least one day ahead of their arrival to prevent surprises for the MC. She works with the planner's input on RSCH jobs including PMs and predictive program tasks and the maintenance coordinator's production needs to best fit work request dates to the available resources for maximum efficiency to satisfy her customers. Her goal is to get the scheduled work completed while handling all emergencies that are coming through the Maintenance Team manager.

The scheduler utilizes planned fill-in jobs to keep her resources loaded when scheduled jobs cannot be done for any given reason. Communication is critical to the smooth teamwork required to pull off the duties of this position. The scheduler works with the other key P&S positions to populate the weekly schedule and then refine the daily schedules. She maintains the P&S measures that track the area's overall performance.

Planner

The planner works to stay ahead of the mechanics and captures value-adding information that is communicated through job packages given to the mechanics as the work is being assigned. He reviews the requested work in the field and defines the scope of the needed repairs, estimates labor-hours to perform (using average mechanic skill level), defines needed support equipment and resources, determines parts and materials required and orders them, and collects support documentation (drawings, pictures, procedures, etc.) that will help the mechanics better understand and perform the work. Once the scope is fully understood, the planner will decide with input from the Maintenance Team manager and others as appropriate if the job is to be sold as a complete turnkey job, if it is to be sold for specific work order operations, or if it is to be performed by area mechanics. This will determine the level of coordination required for communicating with the scheduler. For future work, planner-to-planner discussions are required to coordinate or hand off work tasks. The planner will supply just the right level of information that best fits the needs for each specific job. He will decide when it is best to create a stored plan, task list, detailed procedure, etc.

Material Coordinator

The material coordinator expedites the parts and material for all emergency jobs and then obtains and stages materials for planned work as specified by the planner. Mechanics are expected to get their own parts for emergency needs if it is the most efficient option. If they have value-adding work that can be done while the parts are being obtained, the mechanics will contact the material coordinator to get them. She is the main interface with IMS for obtaining and tracking emergency materials. She has a P-card (purchasing card) for nonstocked purchases. She executes the delivery process to ensure materials are available when and where they can best be used. The material coordinator handles the return of surplus material to stores with proper credit being received. She resolves material questions when the stocked GMN does not fit the mechanic's need. She owns the repairable item process for her area and tracks to resolution to ensure availability when needed.

Maintenance Team Manager/Supervisor

The Maintenance Team manager spends approximately 50% of his time in the field to coach mechanics, to troubleshoot problems, to eliminate delays and roadblocks, and to drive mechanic training needs through assignments and classes. He coordinates the daily work in his area by all mechanics (owner and contractors) to execute the daily schedule while handling emergencies that occur. He observes start of work to final sign-off at completion by checking on the work throughout the day. He uses the findings to improve work processes and to help eliminate delays for the mechanics. He focuses on managing cost savings for labor and materials.

The Maintenance Team manager provides recognition and reinforcement to his people and the operators that support them when they do a good job and make improvements. He encourages mechanics to provide feedback on stored plans and needed changes to BOMs. He initiates identified follow-up actions to prevent future or repeat failures. He leads shutdowns for the mechanics from the field for all his plants. And he carries a beeper or radio to stay in touch and to be available as needed.

• • •

Depending on the size of your organization, one person may have to perform more than one of these roles. The key is to separate the planning and scheduling activities so that they are not done by the same person. Planning is adding value for the mechanic through understanding and preparing a job request prior to the start of tool time. Planning is a future activity. Scheduling is assigning all resources at the optimum time to allow the most efficient execution of a job request. Scheduling is a "now" activity. If the same person is doing both the planning and the scheduling, "now" will always win, and the person will never have time to get out ahead of the mechanics with their planners' job reviews in the field.

Initial Insanity Cure Timeline

The success of these roles is based on the quality of your communication process. This example gives a typical timeline for a very reactive facility to show the needed steps for P&S interactions to stay one day ahead:

- Meet with night crew[*]
- Early look for tomorrow[*†]
- Midafternoon confirmation[*]
- Publish schedule for tomorrow[*]
- Communicate and get "two-day-out wants"[‡§]

Notes

* Meeting of key P&S functions: The maintenance coordinator and scheduler meet with the planner; material coordinator's and team manager's input.

† The "early look" creates the list of work to do tomorrow. It is based on the weekly schedule with urgent need changes included.

‡ Meeting with operations and maintenance: The maintenance coordinator meets with operations; the scheduler meets with the Maintenance Team manager.

§ Meeting with night crew: The maintenance coordinator and maintenance team manager get "true emergency" input for required changes to the day's schedule with minimum disruption of mechanic assignments.

Use the "early look" discussion to complete these tasks:

- Briefly review each job scheduled for tomorrow.
- Answer these questions prior to the midafternoon meeting:
  - Can the equipment be ready (shut down and cleaned up with permits ready)?
  - What resource conflicts exist? Can they be resolved?
  - Where should parts and materials be staged for each job?
- Walk down each job for the MC and the mechanics to verify safety.
- Identify actions for key players to take during the day to be ready for the midafternoon meeting to set the final schedule.

The midafternoon confirmation is a quick check step to verify that the proposed list can be done (equipment can be shut down and prepared tonight including permits and LOTO, parts and materials are available and staged, job packages are ready to hand out to mechanics, and the mechanics with the right skills will be here tomorrow). Mechanics may be owners or contractors.

Use this discussion to complete these tasks:

- Update the status of jobs being done today and incorporate new additions.
- Learn from delays and problems to make the required adjustments to the schedule.

- Determine what to do about carry-over jobs (when to reschedule, put on hold until parts arrive, etc.) and what the impact will be on tomorrow's schedule.

The "Publish schedule for tomorrow" is the communication-sharing step to let everyone know who will be in the area tomorrow and what they will be working on.

All the information must be documented so that it can be efficiently shared with all parties needing to take actions to implement the schedule. Many different software and systems can be used for an appropriate scheduling tool. In the discussion that follows, I use our spreadsheets as an example. The tool that you select must be available to operations, maintenance, and all support resources. It is to show the following information: work order number, short text description, priority of need, ranking of importance relative to the other scheduled work, number of mechanics per job, labor-hour estimates to complete, permits and special preparation needs, and a place to mark work as completed or as carry-over with an explanation at the end of the day. The jobs should be arranged in sections for scheduled work, emergency or unplanned work, fill-in work (when scheduled jobs cannot be performed as shown), and contractor/nonarea resources. The last section is used to tell the operations people who will be coming into their area and what they will be doing to allow equipment preparations and permits to be ready.

The spreadsheet has a place to copy and paste the SAP work order backlog for easily moving jobs from day to day (tabs for each day of the week). Another of the worksheets contains the mechanic roster to capture who is available each day. After adjusting for vacation, sick days, etc., the schedule is then populated to cover 100% of available resources with the jobs ranked to know which lower-priority jobs will be deferred if emergencies occur. This schedule is a mutually agreed-upon list of work based on the interactions of the daily communication process. The P&S measures can then be calculated for the schedule data (daily and rolled up into weekly and monthly averages).

Schedule Performance Metrics

This example is intended to explain the schedule compliance and schedule breaker measure definitions used to drive the initial culture change and to show the progress trends for these P&S process behavior changes. In the daily discussions between the scheduler/Maintenance Team manager and the maintenance coordinator, they decide today to do 10 jobs tomorrow and issue them on the schedule for the next day. At the end of the next day, they score their performance.

Data Captured

Jobs scheduled = 10

Total jobs completed = 20

Scheduled jobs completed = 5

Add-on/break-ins completed = 15

(Emergency jobs completed = 10) Note: E-jobs are a subset of the total break-ins.

Measure/Performance

% schedule compliance = 25% (5 of *20)*
(required P&S Improvement Team measures)

% break-ins = 75% (15 of *20)*
(required P&S Improvement Team measures)

% jobs completed as scheduled = 50% (5 of *10)*
(optional P&S Improvement Team measures)

% emergency jobs of the total jobs completed = 50% (10 of *20)*
(optional P&S Improvement Team measures)

An approved work flow process is the other critical element for selling P&S to the entire site. The example described in Figure 7.2 can be modified to fit your specific needs and organization. (This P&S work process map is available in color to better understand each functional group at www.maintenanceinsanity.com.) *It is offered as a starting point for your team to develop its own work flow to break the insanity cycle.*

The elements of Figure 7.2 include the following.

E-jobs

A fully defined and enforced emergency work criterion with accountability by line management is implemented.

1. Send requests directly to the maintenance supervisor for immediate action.
2. Minimize planner assistance for creating the order, verifying correct functional locations, and sharing available task lists, check sheets, or instructions.

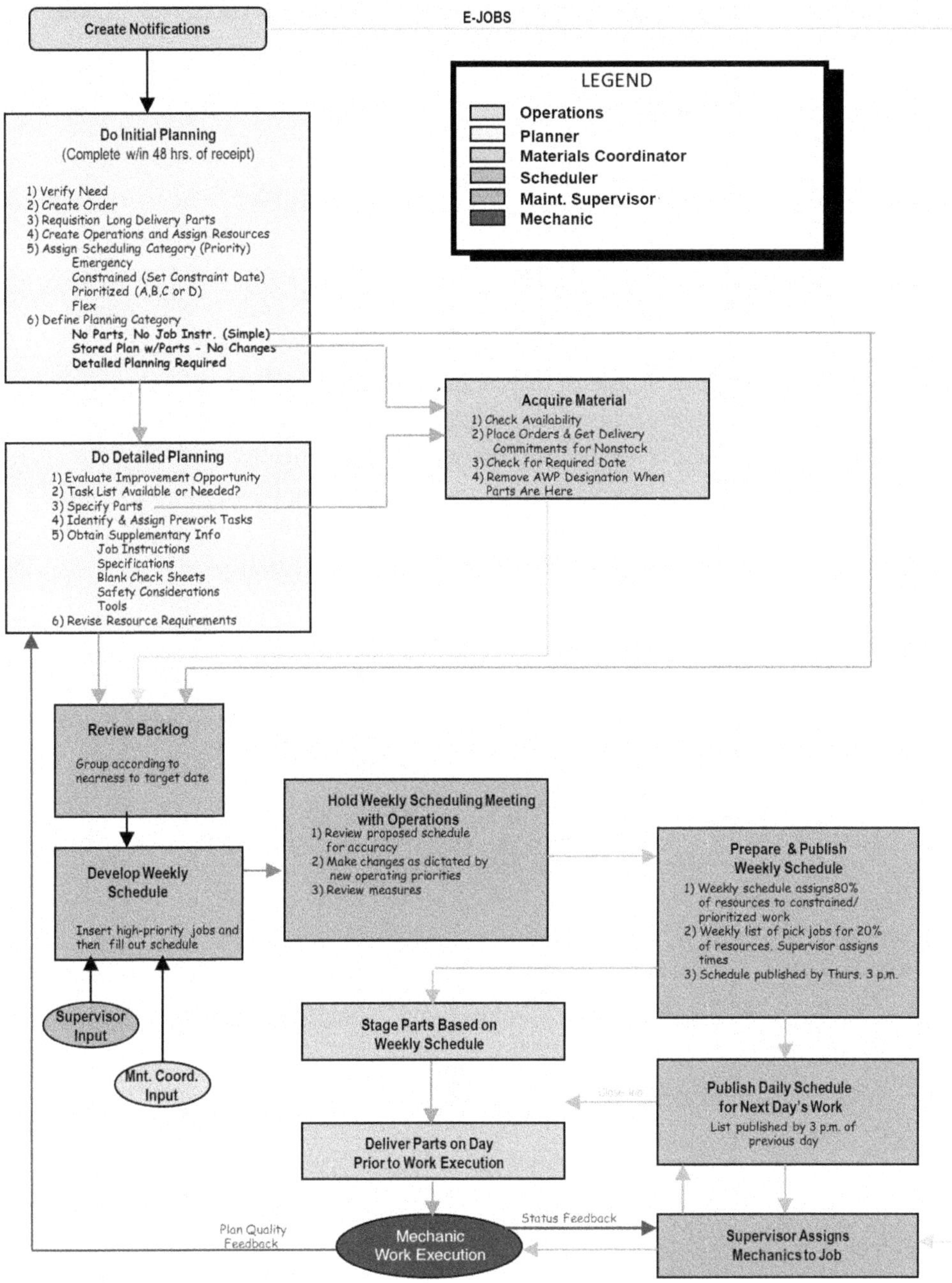

FIGURE 7.2 Planning and scheduling process flow "block" details

3. Tie up loose ends for rotational spares, reporting history and feedback, and closing the work order.

Create Notifications

The maintenance coordinator/production assistant function is to filter identified desired work from all operations, management, and engineering sources.

1. The number of people writing notifications will be restricted to a critical few (maintenance coordinator and shift supervisors). Others will communicate needs to them verbally or through Outlook.
2. The maintenance coordinator or production assistant function will serve as the funnel or filter for operational requests to eliminate duplications and to clearly define what is needed and its appropriate criticality.
3. Required information for each notification will be established and communicated to the users.
4. The use of a site priority system will be enforced, with urgency guidelines to align work by types and level of needs.

Do Initial Planning

The planner function is to be done for all routine nonemergency work. These items are to be completed within 48 hours:

1. Verify need.
2. Critique and correct notification information prior to converting into a work order.
3. Create order.
4. Requisition long delivery parts.
5. Create operations and assign resources (work groups and equipment needs).
6. Pull in the existing task list.
7. Assign the scheduling category (priority).
   a. Emergency.
   b. Constrained (set a constraint date).
   c. Prioritized (A, B, C, or D).
   d. Flex.
   e. Shutdown. (Equipment outage planned by production. Production develops a "Work when down" list, with estimated time to completion, to be

ready for an unexpected outage. Plant outage: revision number and on-site shutdown schedule.)

8. Define the planning category.
   a. No parts, no job instructions needed (simple)
   b. Stored plan with parts—no changes
   c. Detailed planning required

Acquire Material

The material coordinator function is to understand the nature of the work—the equipment being repaired and the impact of each component on the job. Once you have mastered tracking activities on a spreadsheet, then you can define what you will need from an IT tool. This is better than buying some software and changing everything you do to fit it.

1. Be aware of fixed dates versus floating dates to prioritize your actions.
2. Check availability. (For immediate needs, follow up with the requestor and answer questions.)
3. Place orders and get delivery commitments for nonstock.
4. Verify the required date and the ability to deliver. (Expedite to get on time.)
5. Keep the scheduler informed of the status and remove the AWP (awaiting parts) designation when parts arrive.

Do Detailed Planning

The planner function is performed where value can be added to the job being done.

1. Visit the jobsite for field scope refinement.
2. Evaluate possible improvement opportunity.
3. Review and modify the task list or create a new one.
4. Specify parts and write a requisition for materials (designate the people to call in the comment box so that they can be notified of constraints).
5. Identify and assign prework tasks (use work order operations for the other resources required).
6. Obtain supplementary information.
   a. Job instructions
   b. Specifications and standards
   c. Blank check sheets
   d. Safety considerations

   e. Tools required
7. Revise the resource requirements.
8. Set the user status: AWP, HOLD, RSCH, LNC (planning complete).

Review the Backlog

The scheduler function is to work with ready-to-schedule jobs with input from the planner, maintenance supervisor, and maintenance coordinator to set the work schedule for optimum efficiency.

Group the work according to the nearness to the target date.

- Fixed date—cannot move job
- Float—able to move job
- Filler—can do anytime

Develop a Weekly Schedule

The scheduler function populates the work schedule.

1. Insert the highest-priority jobs and the jobs with specific time constraints first.
2. Continue to fill the schedule to achieve 100% resource loading with the agreed-upon percent of filler work.

Hold Weekly Scheduling Meetings with Operations

Face-to-face group interactions are held to update the work request status and to ensure clear communications.

1. Review the proposed schedule for accuracy.
2. Make the necessary adjustments as directed by changes in production demands.
3. Review the performance measures and capture the action items to drive improvements.
4. Finalize the schedule and roll it out to all parties.

Prepare and Publish the Weekly Schedule

The scheduler issues the agreed-upon schedule.

1. The weekly schedule assigns 80% of the resources to constrained or prioritized planned work.
2. The remainder of the schedule is populated with filler jobs.
3. The supervisors assign times and people as the work is to be performed.
4. The weekly schedule is published and available on the web by 3 p.m. each Thursday.

Stage Parts Based on the Weekly Schedule

The material coordinator function will collect the appropriate GMN materials from the job BOMs and needed nonstocked items and package them for staging.

Publish the Daily Schedule for the Next Day's Work

The scheduler issues this work list after final adjustments and confirmation of parts and equipment availability. The daily schedule for next day's work:

1. Provides details for operations about what equipment to prepare during the night shift to have it ready for the start of work the next day.
2. Shows permits and LOTO requirements for each job.
3. Allows mechanics to review the job scope for the next day's work. (The supervisor hands out job packets as assignments are made for the next day.)
4. Communicates all resource needs to the other groups involved.
5. Is issued through Outlook at 3 p.m. daily for tomorrow's schedule.

Deliver Parts on the Day Prior to Work Execution

The material coordinator makes parts available to the mechanics.

1. Materials are staged in the designated locations in the appropriate areas. (The location is shown in the job packets.)
2. Parts are put in bins or boxes for the mechanics to collect for each job.
3. The planner verifies the correct parts (spot-checks), and the mechanics review the contents as part of their job package preparation.

Execute Work as Planned and Assigned by Supervisor
Assigned mechanics to perform jobs per the daily schedule.

1. The job starts at the scheduled time with the proper permits, LOTO, and equipment preparations ready when the mechanics arrive. A quick verification and sign-off allows tool time to begin with minimum delay.
2. The mechanics arrive with the right tools, parts and materials, and support resources to complete the work. They have information (plans and procedures) with them to fully define the scope and to perform the needed repairs as part of their job packet.
3. When the job is completed, the mechanics close the loop with operations by "signing off on" the permits. They provide appropriate feedback to the planner on plan quality and content to improve the job if done again in the future. They update their supervisor on job status, close their operation on the work order, and enter the required technical reports.
4. Unused parts and materials are returned to the material coordinator for restocking in stores.

Check Steps
Audits should be considered at various steps in this process to drive improvements in the work flow and in the established performance measures.

• • •

We will now refine our general flowchart steps to more specifically fit the needs of a small site moving from insanity to a customized P&S environment.

CHAPTER 8

Small Site Planning and Scheduling

Small sites have some unique challenges in the fight against maintenance insanity since they do not have the resources and support organizations that the large sites can justify. Since small sites have to depend on help from the larger sites that may be far away, they learn to wear many hats and become self-sufficient in handling whatever comes their way.

A small site success story is captured in the following processes implemented in a European site by my colleague (Tom Ewing) with great results, including a cultural change that improved the site organization and its strategy and methods for planning and scheduling.

Overview

This small site planning and scheduling strategy has two goals: to increase plant reliability and reduce maintenance cost. Each of the eight key elements of this strategy, when implemented properly, produces one or both of these desired results. However, because there is an inherent cost in the resources required to plan and schedule work, failure to implement the process effectively can actually increase cost. Planning and scheduling every job is not cost-effective. The trick is to use planning and scheduling where it adds value.

The key elements of this strategy are universal. They should be a part of each site's planning and scheduling process. However, the tools and methods that are used may vary depending on the size and other unique characteristics of each site.

Finally, efficiencies gained in planning and scheduling do not go to the bottom line unless the savings is actually removed from the maintenance cost structure by reducing external resources, redeploying internal resources, or reducing material purchases.

Key Element 1: Classification of Work

Incoming work should be categorized into one of four categories (operations work, emergency work, routine work, and shutdown work) that define the way that the work will be processed in the maintenance system. These four categories, shown in Figure 8.1, give a "big-picture" overview. Additional details are provided for this small site work process at www.maintenanceinsanity.com to ensure you have the detailed information to let you apply these processes at your site.

First Line Work

First line work includes all work that either cannot be planned or scheduled due to time constraints or does not benefit from planning and scheduling due to its simplicity. There are two subcategories of first line work: work managed by operations and work managed by maintenance.

When operations manages the work, maintenance overhead is minimized, communication lines are shorter, and the response can be quicker. However, this type of work must be limited to jobs the operators have been specifically trained to do. To increase the amount of work that can be managed by operations, the operators can complete a self-paced training program developed by the company or vendors, and/or the maintenance department can assign a mechanic or technician to work under the supervision of the operating department. The maintenance technician can then complete a wide variety of work with assistance from the operators.

When maintenance manages the first line work, there is the additional cost of maintenance supervision and SAP processing. However, for large jobs or jobs requiring special skills, maintenance supervision of the work is the only option.

Second Line Work

Second line work includes all work that does not require an immediate response, that has the potential to benefit from planning and scheduling, and that is not dependent on a large unit shutdown. Usually this category includes the majority of preventive and predictive maintenance tasks, alterations, small capital jobs done by maintenance, and equipment repairs not requiring an immediate response.

The primary difference in the approach to second line work (compared with first line work) is that the nonemergency nature of the work provides the opportunity for the work to be planned and executed in an orderly and presumably more cost-effective

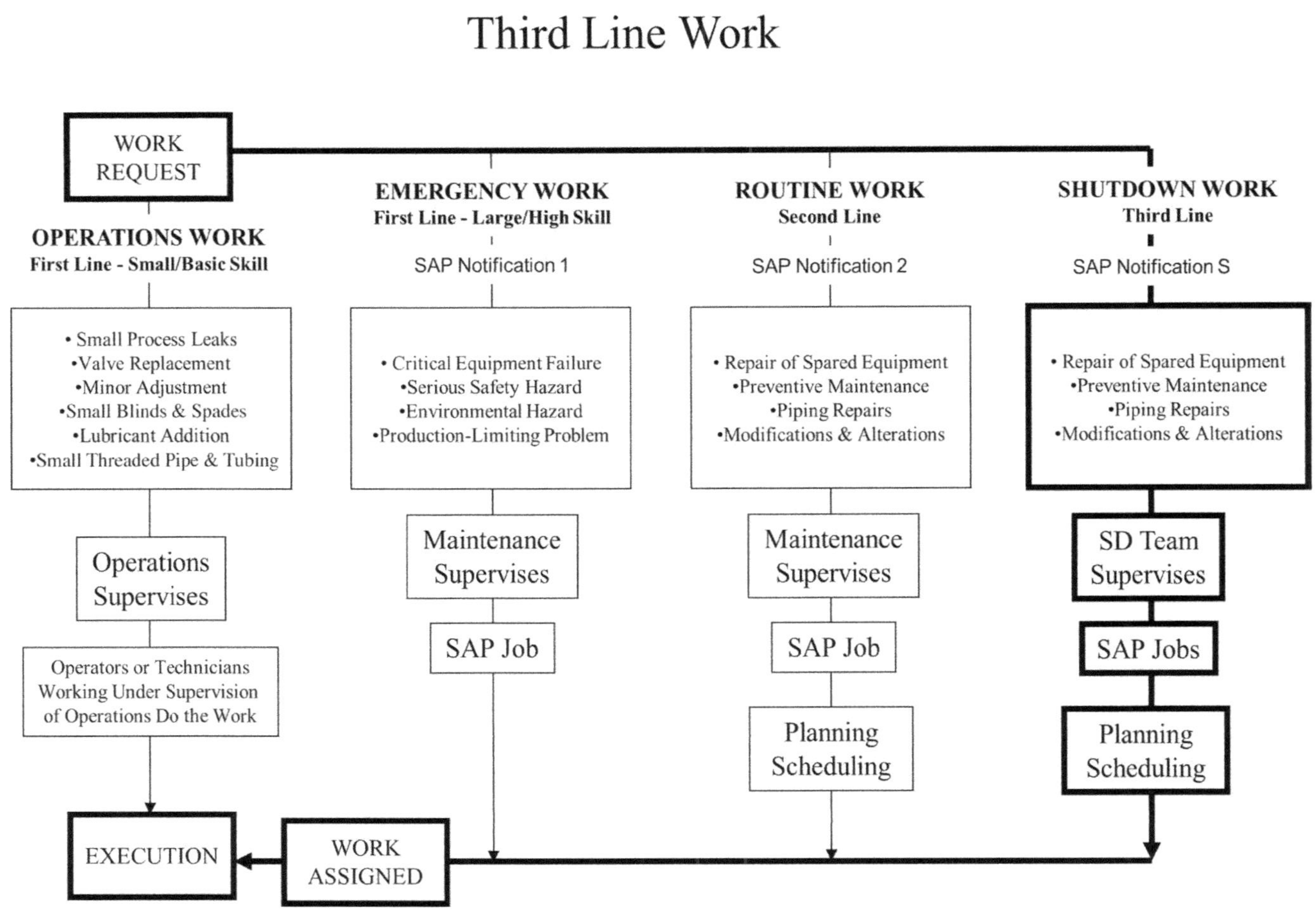

Figure 8.1 Work flow process

manner. However, not all jobs in this category benefit from planning, so each job must be evaluated to determine if the added value of planning the job exceeds the cost of the planning resources.

Second line work is managed by the maintenance department and executed by the site maintenance personnel and/or contractors.

Third Line Work

Third line work is identical to second line work except that the work must be done in conjunction with a shutdown of all or a part of the operating unit. Usually the management of third line work is handled by a temporary organization set up specifically to manage the unit shutdown. However, the identification of the work and planning may occur several months before the shutdown.

Third line work is managed by the shutdown team and executed by site maintenance personnel and/or contractors.

The strategy for classifying work should be:

- Maximize first line–operations for as many of the simple, small jobs as possible.
- Minimize first line–maintenance, using it only when a fast response is justified by business conditions or environmental or safety hazards. Ask operations to consider a second line response if you do not see the business need for an emergency response.
- Maximize second line for as much of the remaining work as possible.
- Use third line only when the opportunity to do the work is in conjunction with a planned shutdown of the unit.

Key Element 2: Maintenance/Production Organization

The maintenance and production departments must be organized to promote good communication, planning, and teamwork during the execution of work. In addition, within the production organization there must be the capability to complete first line–operations work. Experience has proven that an organization similar to that shown in Figure 8.2 can produce good results. However, the size of the site will dictate the options that are available. The principles of direct lines of communication, a "funnel" for maintenance requests, and a method for completing small jobs using operators should be maintained regardless of the organization selected.

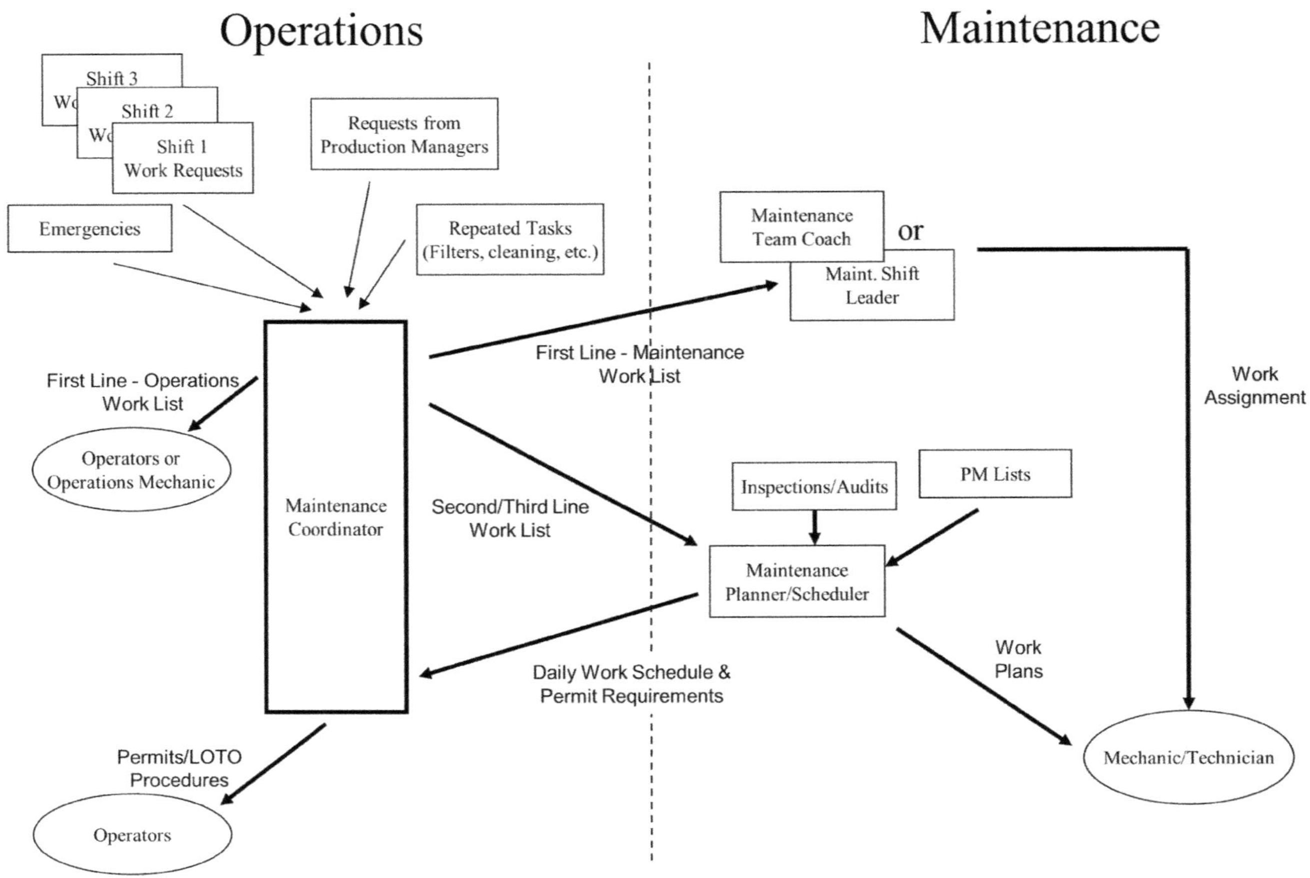

FIGURE 8.2 Operations and maintenance organization

In the organizational structure shown in Figure 8.2, the maintenance coordinator, maintenance planner, and maintenance team coach form the core team that plans and executes maintenance work. Work requests come into the team from various sources, but the team determines which category of work each job fits best, what the priority will be, when the jobs can be scheduled, and what preparations are needed before the job can begin.

At smaller sites, the same person may hold the roles of both team coach and planner. When trying to determine whether a separate planner or scheduler is needed, the following guidelines can be used:

> If there are at least 8 craftspeople, making one of them a planner/scheduler should add value. If there are more than 15 or 20 craftspeople, two planners are recommended. For most small sites, the person doing the planning can also do the scheduling, but if separate planners and schedulers are justified, a scheduler should be able to support up to four planners.

As mentioned in the previous section, the ability of the operating department to do first line maintenance is significantly enhanced by assigning a mechanic from the maintenance department to work directly with operations on first line jobs. The mechanic should remain in the maintenance organization "on paper" and should participate in the same training and development programs as the other mechanics. However, the mechanic should get his or her daily work assignments from the maintenance coordinator and should consider the control room, and not the maintenance shop, as the primary starting point each day.

The strategy for organizing the departments should be:

- Provide a "point person" in the operations department to receive all maintenance requests and determine which ones can be done by operations and which must be sent to maintenance. This point person should also coordinate the maintenance work done by operators and ensure that the equipment shown on the daily work schedule is ready for maintenance at the designated time.
- Designate a point person in maintenance to work closely with the maintenance coordinator. Usually this will be the maintenance planner.
- Appoint a maintenance mechanic to work directly with operations on first line work. This mechanic should remain in the maintenance organization "on paper" but receive the daily work assignments from the maintenance coordinator.

- Establish weekly planning meetings and hold daily discussions to discuss job requirements, progress, changes in the plan, etc.

Key Element 3: Priority System

Within each work classification (first line, second line, third line), there should be a process for prioritizing the jobs so that the most important work receives the proper focus.

For first line work done by operations, the operating department will determine the priority of the work and will maintain a priority system of its choosing. Normally, this will not be an SAP-based priority system. Some of the work done by operations will be urgent, and other work will be routine.

By definition, first line work done by maintenance is urgent, critical work. This level of response is designated by a Priority 1 on both the SAP work notification and the SAP work order. The selection of a Priority 1 indicates that the job justifies overtime, extra shifts, or additional contractors in order to complete the job as quickly as possible. Because there is additional cost for this level of response, and because this type of work breaks into and disrupts the planned work, great effort must be made to minimize Priority 1 work. Excessive use of Priority 1 is usually the primary saboteur of any prioritization system. It is important to measure the amount of Priority 1 work requested and follow up frequently with operations to review the Priority 1 jobs to be sure they indeed deserve an urgent response.

Second line work includes a wide variety of jobs—some that require a relatively quick response and others that may take several weeks of preparation. When an SAP work notification is written, all second line work is designated by a Priority 2 on the notification. However, when the maintenance planner begins to organize the job, he or she needs a way to show if the job must be started within a few days or a few weeks. For this reason, when a notification is converted into a work order in SAP, the maintenance planner has the option of choosing one of four priorities for second line work—A, B, C, or D. Figure 8.3 shows the average target response time for each of these categories. It also shows the type of work typically assigned to each group.

Third line maintenance work (shutdowns) is assigned a priority S in both the SAP work notification and the work order. The shutdown planner will determine the ultimate priority of shutdown jobs.

The selection of Priority A, B, C, or D is an important one and should be made by the maintenance planner in close communication with the operations maintenance coordinator. There is a natural tendency to give all jobs a "middle-of-the-road" priority like

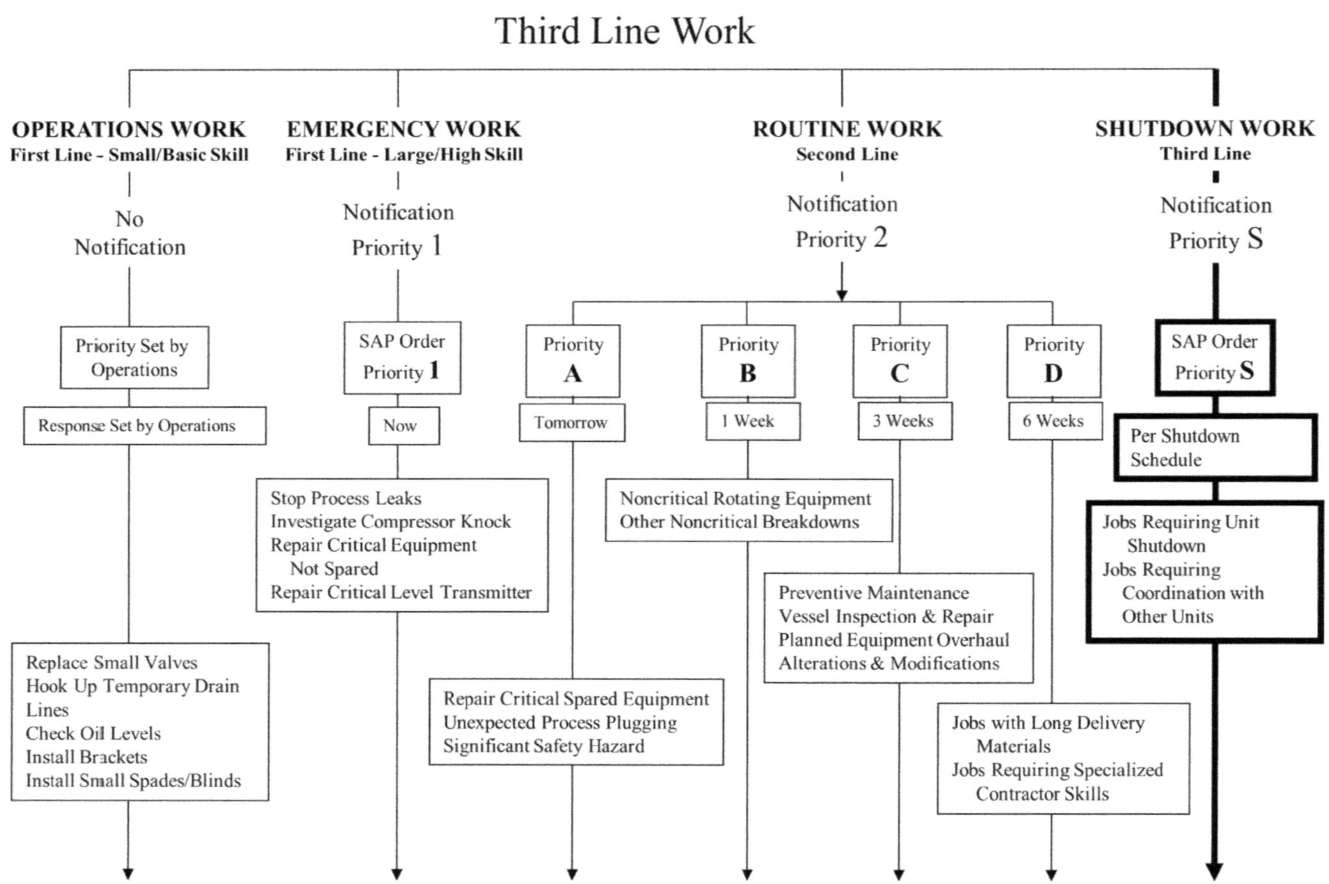

FIGURE 8.3 Average response time and group work assignments

Priority C, but this renders the priority system useless. The maintenance coordinator can provide an initial target date by entering a specific start date in the SAP work notification when he or she creates it, or by typing the information in the long text box at the bottom of the notification. The planner can then use this target date to set the correct priority.

The strategy for prioritizing work should be

- Use Priority 1 only for critical, urgent work that justifies overtime, double shifts, or other additional resources. Measure the number of Priority 1 jobs written, and regularly review these with operations. The number of Priority 1 jobs should not exceed 10% of the total number of jobs assigned to the maintenance department.
- Use categories A, B, C, and D to prioritize second line work. Whenever possible, assign work to the B or C categories to allow adequate lead time to plan and schedule work. If the work involves more than one craftsperson or special equipment such as mobile cranes, it should be given a B or C priority.

A good balance of priorities is 10% Priority 1, 10% Priority A, 30% Priority B, 30% Priority C, and 20% Priority D.

Key Element 4: A Value-Adding Planning Process

A good definition of planning is "providing all the necessary materials, information, and resources to do the job correctly." Simply, it answers the question, *What* is the best way to do the job? Planning a job ahead of time does not automatically add value. For example, I do not need to hire a person to plan the job of mowing my grass. I would not save time or money by paying someone to write down the instructions for operating the mower, and to go to my garage to get the petrol can and place it by the mower a day in advance, and to measure the grass and determine the optimal height of the mower. I can do all those things myself just as quickly and correctly at the time I decide to do the job. However, if I decide to build a storage building in the back of my garden, I would be foolish to walk into the garden to begin that job without having already created some basic design, purchased my materials, and rented the necessary tools. For this reason, each job must be evaluated to determine the amount of planning that will add value to the job.

Planner time percentages for a typical day are estimated as:

Job screening, 5%

Job requirements, 10%

Job research, 5%

Detailed job planning, 20%

Job preparation, 5%

Procurement, 20%

Job scheduling support, 15%

Daily E-job support, 5%

Job closeout, 5%

Personal and miscellaneous, 10%

This sums to 100% for the day.

Not all jobs require planning, but they all require evaluation and field review (no plan, simple plan, or detailed plan). Not all jobs deserve a separate work request. Planners can only plan an average of four jobs per day. Some jobs take several days to plan properly, while others require minimal time.

The basic steps of this evaluation process, shown in Figure 8.4, are minimize, categorize, and optimize.

Minimize

The best possible outcome of the planning process is to determine that the work is not necessary or that it can be significantly reduced from its original scope. Examples of this are knowing that an upcoming capital project will replace a pump that is on the list to be painted or combining several jobs for one contractor to avoid multiple service calls.

In some cases, it is not possible to avoid the job, but it is possible to make changes in the design of the system to prevent future failures. Examples of this are upgrading certain materials, changing the route of a section of pipe to eliminate a restriction, and installing an entry point into a system to allow online cleaning. This step has the most potential for real savings but is often ignored in the planning process.

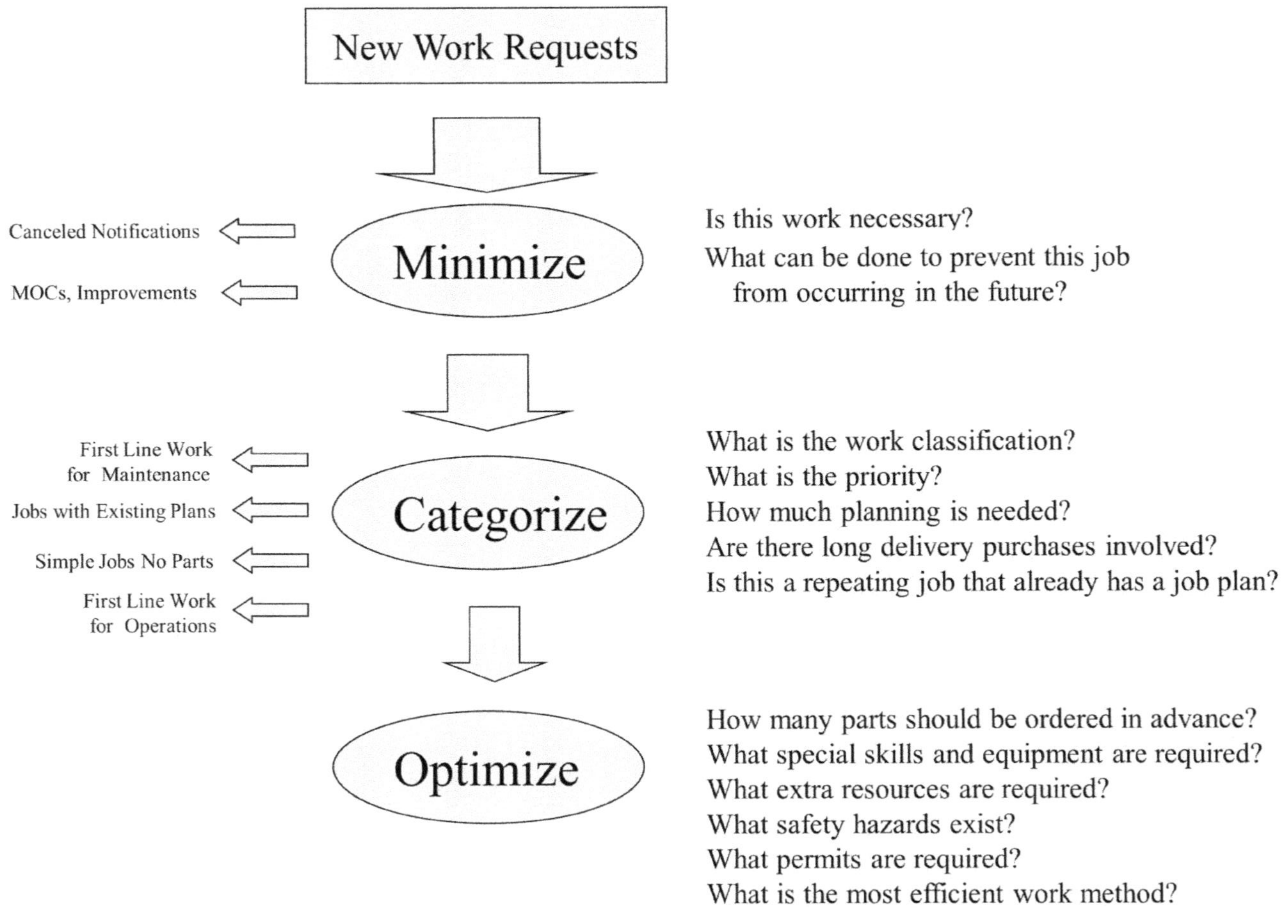

FIGURE 8.4 Work evaluation process

Categorize (Preplanning)

If the job cannot be eliminated, a preliminary evaluation should take place before detailed planning is done. The preplanning step should occur very soon after the request is received. Then if the job has an impending due date or requires acquisition of materials with long delivery times, the planning can be put on a fast track. This is also the time where the work is assigned to one of the work classifications and priorities mentioned above. Finally, the categorization step is the point where the level of planning and scheduling needed to add value is determined.

Optimize

Once the job has been categorized, the detailed planning can begin. As mentioned previously, the amount of planning that adds value will vary from one job to another. A planning "package" should be developed for each job; the contents of the package may be as simple as an SAP job ticket or as complex as a stack of documents including a parts list, detailed work instructions, vendor drawings, LOTO instructions, and MSDS sheets. In general, the following four categories should be addressed in a job plan:

1. **Resources.** In order for the job to be scheduled, it is necessary to estimate the number of people required to do the job and the number of hours required to complete the job. This information should be entered into the scheduling program (SAP, MS Project or Primavera—see Key Element 6: "Planning and Scheduling Tools"). If external resources (contractors) are required, they should also be specified in the plan, and the contractors should be notified of the upcoming job. If special skills or knowledge is required, it should also be noted.
2. **Materials.** A list of the materials needed to do the job should be prepared. If the materials are kept on-site, the stock should be checked to make sure the parts are available. If the materials are not kept on-site, they should be ordered with sufficient lead time to make sure they are on-site before work begins. When practical, the parts should be booked out of stores, delivered to the maintenance shop or jobsite, and tagged with the job number prior to the start of work.
3. **Services.** If mobile cranes, special tools, air compressors, portable generators, hydroblasting, or other services will be needed, these should be spec-

ified in the plan, and the appropriate service provider should be contacted in advance to make a reservation for the needed services.

4. **Information.** Depending on the hazards and complexity of the job, safety advice, MSDS sheets, drawings, sketches, vendor repair instructions, photographs, and other information may be of value. The planner must decide what information will add sufficient value to justify the time required to compile the information.

The SAP job ticket should serve as the basic planning document, and a job ticket should be prepared for each job. The text field of the job ticket can be used to identify safety hazards, provide the location and general instructions for the work, and include a short parts list. In many cases this will be all that is required. If additional documents such as drawings, complete parts BOMs, checklists, or detailed repair procedures are needed, they can be stapled to the job ticket.

The strategy for planning work should be

- Look at the work requests carefully to be sure the work is essential. Make sure that the work is assigned to the work group that can do it most effectively. Identify improvement projects or material changes that will keep the work from having to be done in the future.
- Plan the resources for the job with the minimum number of resources that can do the job safely and effectively. Estimate durations assuming the job will proceed according to plan. If special skills or knowledge is required, be sure to specify this and notify the provider of the special skills.
- Determine the materials needed and place orders in advance for items not in stock. Where practical, purchase parts in advance and deliver them to the shop or jobsite prior to the start of work. Inspect the parts to be sure they are fit for use.
- Determine the services (cranes, insulation removal, special tools) needed for the job and notify the provider of an approximate date the services will be needed.
- Determine the amount of information that should be provided to the person doing the work, particularly if the information is not readily available or commonly known.

Key Element 5: A Value-Adding Scheduling Process

If planning answers the question "how," then scheduling answers the question "when." In a reactive environment with plenty of resources available to handle peak loads, scheduling provides minimal value. However, the implementation of a good scheduling process and a good prioritization process will allow a maintenance department to move from a reactive mode to an organized, forward-thinking mode. In addition, the efficiencies gained should allow the same amount of work to be done with fewer people.

It is critical that the scheduling process be jointly owned and managed by both the production and maintenance departments. Neither group alone has all the knowledge required to create the optimal schedule. Production should provide information on optimal timing to meet production requirements, and maintenance should provide information on optimal timing to meet worker and material requirements.

There is no possibility of scheduling emergency work, and there is usually no value in scheduling other first line work, so the scheduling process applies primarily to second and third line work. In order to schedule effectively, you must have enough lead time to arrange for all the required materials and resources to be provided prior to the start of the job. For a simple job involving only one of your company employees and a few parts kept in stock, only a day or two of lead time is required to schedule effectively. However, if external resources such as cleaning contractors or specialty repair services are involved, at least a week of lead time is desirable. If the production departments are unable to provide this lead time, you should work with the production department management too.

For this site, two schedules are recommended: The first is a weekly schedule showing all the jobs that are expected to be executed in the upcoming week. The second is a daily schedule that is provided to operations on the evening prior to the day's work showing the jobs that need to be prepared for execution the next day.

The weekly schedule should be prepared midweek of the previous week and reviewed with operations toward the end of the week, preferably in a weekly production/maintenance meeting. The weekly schedule should contain enough work to allocate approximately 80% of the available second line resources for the following week, leaving the remaining 20% available for high-priority second line work that can be expected to occur during the week. This relationship is shown in Figure 8.5 in the form of a sample schedule.

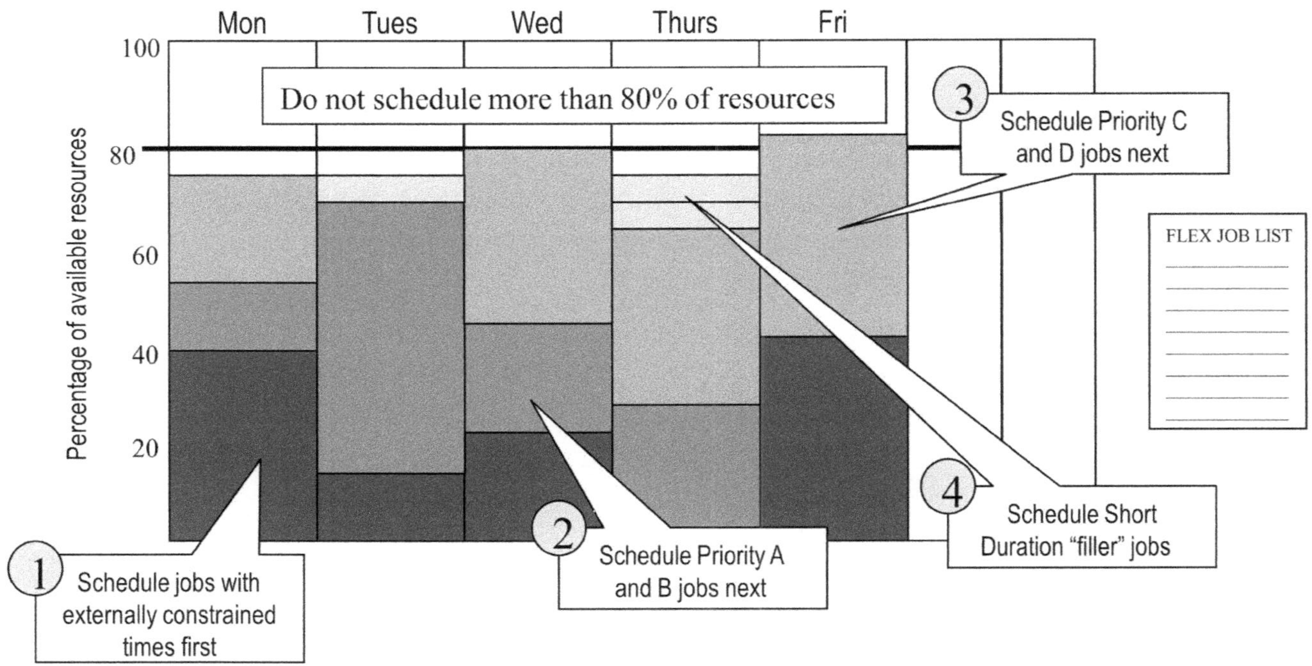

FIGURE 8.5 Sample schedule

The weekly schedule should be built in the following order:

1. Begin by scheduling jobs with external constraints that force execution on a specific date; constraints include availability of government inspectors, timing of production changeovers, availability of contractors or vendor consultants, etc.
2. In the remaining open spaces, schedule all Priority A and B jobs that will fit into the available windows.
3. If additional scheduling space remains, schedule all priority C and D jobs that will fit into the available windows.
4. Normally following this process, a few short-duration windows will remain scattered throughout the schedule. If more than 20% of free resources still remain, look through the backlog and attempt to fill these gaps with short-duration jobs.
5. Finally, prepare a list of 5 to 10 relatively small, noncritical jobs that can be done with minimal preparation by operations and involve only company maintenance. These "flexible" jobs will be executed if the expected 20% of new second line work does not materialize or if the scheduled jobs are postponed at the last minute. Then the team coach can assign one or more of these flexible jobs for immediate completion.

The primary purpose of the weekly schedule is to provide a communication tool for operations and maintenance to agree on the general package of work to be executed during the following week. In the weekly planning meetings, the next week's schedule should be reviewed, and all parties should agree that (1) the equipment can be made available for work and (2) the required parts and labor are ready. However, in most cases a few changes in the schedule will occur due to unexpected failures or changes in plan. For this reason, a simple daily schedule prepared each afternoon for the following day's work is also needed.

The daily schedule is simply a list of the jobs to be done the following day, with a list of permits required and any special preparation instructions. If possible, the schedule should also include the names of the mechanics and technicians who will be doing the work. This schedule should be prepared by the maintenance planner and e-mailed or delivered to the operations shift leader and maintenance coordinator late in the afternoon. If a job is shown on the daily schedule, it is assumed that maintenance is definitely planning to do the work, and operations can prepare the work on the night shift unless otherwise noted.

At this site, we selected Microsoft Project as our scheduling tool. MS Project will produce both a weekly and a daily schedule in a few simple steps. These are described in more detail in the next section.

The strategy for scheduling work should be

- Schedule only second and third line work. Work with operations to ensure that there is sufficient lead time to plan the work and organize the needed materials and resources.
- Develop both a weekly and a daily maintenance schedule. The weekly schedule should be prepared midweek for the upcoming week and include 80% of the available second line resources. It should be prepared with the help of the operations maintenance coordinator. The weekly schedule should be completed prior to the weekly production/maintenance meeting and reviewed at that meeting.
- To create the weekly schedule, insert the externally constrained jobs first, followed by Priority A and B jobs that are ready to be executed. Then fill the remaining work windows with Priority C and D jobs. For the remaining small windows, insert short-duration jobs of any priority.
- Prepare the daily schedule the day before the work and include a list of the work to be done, the resources needed to do the work, and the permits required. The daily schedule should be e-mailed to the shift leader and maintenance coordinator each afternoon.
- Create a list of 5 to 10 flexible jobs that can be completed with minimal operations preparation if scheduled jobs must be postponed.

Key Element 6: Planning and Scheduling Tools

In the European region, we selected SAP as our primary tool for job planning and Microsoft Project as our primary tool for daily maintenance scheduling. MS Project is also recommended as the shutdown scheduling tool for most sites. A few of the larger sites may prefer to use Primavera, which has the ability to handle larger shutdowns but is more difficult to use.

SAP for Planning

By definition, *first line* work does not require planning or scheduling. And for first line maintenance work done by operations, work notifications do not normally need to be written for each job. However, arrangements must be made in SAP to collect maintenance charges. If operators do the work, their labor is not charged to maintenance jobs, and only a "standing order" for material purchases is needed. This standing order is created at the beginning of the year by maintenance for use by operations to buy small miscellaneous parts and materials from stores. If a maintenance mechanic or technician is also permanently designated to assist with first line operations work, the maintenance mechanic (or technician) will need an additional standing order to charge for labor when working on routine jobs not involving critical equipment. If the maintenance mechanic assigned to work with operations works on a larger job involving a critical piece of equipment, that mechanic should write a notification and create a work order so that the history and cost can be booked separately.

For most *second line* and *third line* jobs, a notification should be written by operations to request the work. Second line work should be given a priority of 2 (nonemergency) or S (shutdown required) on the notification. If a specific execution date is desired, the person writing the notification should record it in the "Requested Start" field. Details about the job—a brief description of the problem, the person who can provide additional details, and any special requirements—should be recorded in the text box at the bottom of the notification. If the job is the result of an equipment breakdown, the "Breakdown" box should be checked. The correct functional location should be entered because the functional location determines which work group will receive the notification.

The maintenance planner should check SAP for new notifications at least twice daily and contact operations to discuss any urgent jobs. For other jobs, the maintenance planner should complete the planning process as explained above. Notifications should be converted to work orders as soon as it has been determined that the job is necessary. Normally, the work order should also be released for charges at this time unless there is a specific reason that charges need to be blocked for a time. Basic instructions can be typed into SAP using the long text field, and if the site has fully implemented SAP, parts can be ordered from stores using a bill of material attached to the functional location. The SAP job ticket can be printed for each job. Additional information such as drawings, sketches, MSDS sheets, etc., can be stapled to the job ticket.

In addition to notifications and work orders, SAP has the ability to break the work order into smaller subtasks called operations. However, for this site we are not including the use in our strategy because it is generally easier to do in MS Project than in SAP. However, if sites prefer to use SAP operations as a means of designating work for contractors or for other means, they may do so.

MS Project for Scheduling

Only second and third line work is normally scheduled. By definition, first line work does not benefit from planning or scheduling. Scheduling daily work in MS Project is a four-step process:

1. Import the jobs to be scheduled from SAP.
2. Assign a start date, duration, and crew size.
3. Assign subtasks if desired.
4. Assign the names of the people who will do the work.

The maintenance planner/scheduler is responsible for building the schedule in MS Project, but it is possible to allow others such as a maintenance coordinator or contractor planner to modify the schedule or add information. In addition to the basic scheduling information (dates, resources, durations, etc.), MS Project has several places where notes, links to other documents, etc., can be added.

It is possible to send copies of the various MS Project schedules and reports to a wide distribution using MS Outlook e-mails. Although only weekly and daily schedules are normally printed, MS project can be thought of as a continually changing and growing calendar that contains all jobs that need to be tentatively scheduled, even several months in the future. If dates change, they can be easily updated in MS Project. Completed jobs can be easily removed.

MS Project can also be used to track the status of multiple day jobs. This site's strategy does not demand that this status feature be used, but it is available if a site finds it desirable. Jobs that are complete can easily be sorted in MS Project and the list of jobs for closure copied quickly to SAP for confirmation (closing).

In order to use MS Project for scheduling, it is necessary to obtain more detailed training than can be included in this document. However, MS Project is user friendly and intuitive and has numerous features that can allow more advanced scheduling (resource leveling, etc.) than those outlined in this strategy.

Shutdown Scheduling Using MS Project

The procedure for scheduling a shutdown using MS Project is basically the same as scheduling for daily maintenance, with the exception that the resource-leveling features of the software may be beneficial, and it may be useful to arrange the resources into different groups.

Additional information on using MS Project for shutdown scheduling will be added to this document at a later date.

The strategy for using planning and scheduling tools should be

- Use SAP for planning.
- Use MS Project for daily scheduling. Prepare both a weekly and a daily schedule and review the weekly schedule in a weekly planning meeting with operations.
- Use MS Project or Primavera for shutdown scheduling.

Key Element 7: Measures

The key to successfully implementing any new process is a good implementation plan and a good follow-up process that monitors key indicators. A good monitoring process will identify areas that need additional emphasis, areas that need to be changed, and areas that need to be positively reinforced. Several key measures for planning and scheduling give an indication of the health of the planning and scheduling process. The suggested measures for this site's planning and scheduling are detailed below.

Percentage of Emergency Jobs Completed by Maintenance (Priority 1)

This is a measure of the amount of reactive work required. Reactive work is generally less efficient and interrupts the scheduled work. Therefore, it should be minimized. As an initial goal, a site should set a target of 50% of its current reactive level. Once this level is reached, a final target of 10% should be set. This measure can be obtained from SAP with minimal effort.

Percentage of Total Maintenance Hours That Are Done as First Line Work by Operators

This is a measure of the level of involvement of operations in doing first line maintenance. The target for this measure should vary depending on the type of organization that is set up (mechanic assisting or not, number of operators participating, etc.), but the number should probably be between 10% and 25%. This is a valuable measure, but there is one big disadvantage—it must be manually recorded by operations; there is no way to obtain it automatically from SAP.

Percentage of Jobs in Each of the Four Second Line Priority Groups—A, B, C, and D

This is a measure of adequate lead time. Approximately 70% of all second line work should fall into category B or C, approximately 10% in category A, and approximately 20% in category D. A very high percentage of category C work can indicate a tendency to give all jobs an "average" priority of C, which means that the priority system is not really being used. This measure can be obtained from SAP with minimal manual effort.

Percentage of Jobs Planned

This is a measure of planning activity. However, remember that not all jobs should be planned—and the level of planning will vary from job to job. These factors make it almost impossible to create a good measure for planning effectiveness. Probably the best approach is to ask the mechanics to put completed job plan packages in a specific location, and then you audit approximately 10 packages per month, using a set of specific criteria (checklist) to score the effectiveness of the plan. A second, less useful approach is to count the percentage of plannable jobs that have received at least some minimal level of planning (basic job instructions and resource/material requirements). A goal of 80% of plannable jobs is a reasonable initial target. This measure can be obtained from SAP if the "Planning" check box on the "Additional Data" tab of the work order is checked at the time the planning is done.

Percentage of Total Hours Worked That Were Scheduled

This is a measure of how much of the work done was included in the weekly schedule. An initial goal of 60% is suggested, with a final target of 80%. The 80% is derived by assuming that there will be 10% unscheduled work due to emergency jobs and another 10% due to completion of jobs on the "flexible" list that are intentionally left out of the schedule to provide some flexibility for the team leader. This measure is more difficult than most of the others to obtain because it requires retrieval of the hours worked from SAP and the hours scheduled from MS Project. We will attempt to provide templates or customized reports to make this measure less difficult to create.

Percentage of Scheduled Jobs Completed

This measure sounds like the previous measure, but whereas the previous measure monitors how much of the total workload is scheduled, this measure monitors how well we do at completing the work that is scheduled. If we schedule almost all our work but don't complete any of it on time, the first scheduling measure will be good, but the second one will be bad. Conversely, if we complete all our scheduled work but only scheduled two of twenty total jobs, the second measure will look good but the first one will look bad. So both measures must be used together. Since the schedule only includes second line work, and since we are scheduling just 80% of the available second line resources, this measure should approach 100%. A good initial goal would be 80%, with a move to a 100% target after several months.

• • •

The strategy for measures should be:

- Measure only a few key factors and set up feedback channels with operations to review and take action when they do not meet the target. Set an initial target and gradually increase the target as you become better at planning and scheduling. Track the following measures:
  - Percentage of emergency jobs completed by maintenance
  - Percentage of total maintenance hours that are done as first line work by operators
  - Percentage of jobs in each of the four second line priority groups—A, B, C, and D
  - Percentage of jobs planned
  - Percentage of total hours worked that were scheduled
  - Percentage of scheduled jobs completed

Key Element 8: Process for Taking Advantage of Efficiencies Gained

Implementation of this planning and scheduling strategy should improve the quality of repairs, reduce waste, and, perhaps most prominently, eliminate delays and miscommunication that reduce wrench time. As wrench time increases, resources are freed for other work. At the same time if the planning process is effectively designing out problems, the workload should go down as well.

If cost is to be removed from the maintenance system, these newly free resources must either be used to reduce external resources (contractors) or be assigned to productive work outside the maintenance budget. Some options to consider are

1. Assume responsibility for functions done previously by contractors if your employees can do these functions effectively.
2. When openings in operations shifts arise, fill them with maintenance mechanics. This option is especially attractive when some of the more mature mechanics have requested a challenging job that takes advantage of their extensive experience but exposes them to less of the hard physical work that some maintenance jobs entail.
3. Cross-train mechanics to do both electrical and mechanical work.
4. Provide mechanics with opportunities to move into support groups as engineering assistants, IT technicians, or other nontraditional technical jobs.

The strategy for taking advantage of efficiencies gained should be to:

- Look for opportunities to eliminate or minimize the use of external resources.
- Give experienced mechanics the opportunity to move into operations when openings occur.
- Consider nontraditional technical roles such as engineering or IT.

You can see from the level of detail that even a small site needs a solid structure with common understanding by all parties to operate efficiently and break out of the insanity "do-loop" of just keeping your head above water. This process proved to be very successful in turning this small site around and into a benchmark location.

The next chapter will address costs and how to analyze them to see how much your maintenance insanity is putting you in the hole.

CHAPTER 9

Cost Analysis

Cost Analysis and Control

One guaranteed way to stay stuck in maintenance insanity (doing the same things and expecting different results) is to not measure anything or the right things. You do not know how much your inefficiencies are costing until you capture and analyze them. Here is a simple method to help you get a handle on your spending and understand where the costs are going.

Capture cost in these six buckets:

1. **Internal labor.** Your mechanic hours paid (part of full-time equivalent calculation and includes overtime)
2. **External (contractor) labor.** Contractors charging by the hour (part of FTE calculation and includes their overtime)
3. **Internal materials.** Items bought from your stores (part numbers set up for your site or used from another site)
4. **External materials.** Purchase order amounts paid for M&R supplies, parts, and materials not set up in stores
5. **Internal services.** Overhead for supervision, staff, and other support and for accounting adjustments
6. **External services.** Third-party services not based on hours worked (lump sum, annual fees, etc.)

There is no need for a miscellaneous bucket (it confuses evaluation and serves as a catchall). Please work to eliminate this category if you have it now. These charges to miscellaneous need to be assigned to one of the other six buckets. A graphic representation of the six-bucket monthly tracking is shown in Figure 9.1, and the entire spreadsheet is provided at www.maintenanceinsanity.com for your use.

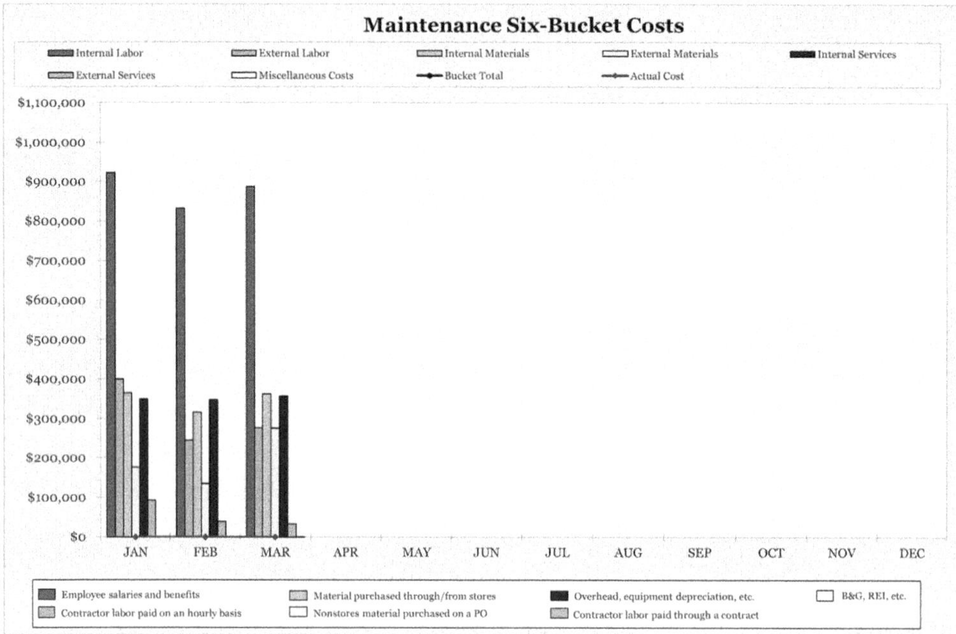

FIGURE 9.1 Six-bucket comparison

General Rules of Thumb Pertaining to These Buckets

If mechanics are efficiently focused on making repairs and improvements, there will be an appropriate 50-50 split between labor and material charges. If you cut labor, the material costs should fall about the same amount. If your ratio is higher on the labor portion, you are probably spending a lot of time on nonmaintenance activities (unplugging, unchocking, operating activities, etc.). Check your ratio and investigate how to increase tool time (planning and scheduling improvements, transferring operating duties to operations, etc.).

Once you have established a storeroom, the internal to external material should be in the 70 to 30 range to keep from having to hustle materials on a daily basis. It can be different if you have vendor stocking agreements where the vendors hold parts and materials for you and you only pay when used. We need to limit our inventories but not be caught by surprises routinely (lost production because critical or long-term parts were not available).

The ratio of your labor to contractor cost should be driven by the need for site core skills and overall cost to perform the needed service. Contractors can be a lower-cost value or maintain a special skill (not frequently needed). Our mechanics must be fully utilized and have the option of displacing contractors when appropriate for optimum resource utilization of contractors. This can be accomplished through centralized planning and scheduling of capital and expense activities for the site. Good contractor management processes are also key to this success.

Contractor labor and external services add together to give your total outsourced cost. External services should not be affected by the number of people or hours worked. This should be a firm bid and negotiated cost per job activities. If your system setup is lumping all contractor costs into this one bucket, you still need to manage them as separate functions.

Cost Analyst Duties

It always helps to have an owner for any function. The following is a list of expectations for a cost analyst position for direct and indirect procurement to manage cost analysis:

1. Develop an invoice verification process.
   a. Manage the interface with accounts payable in central services.
   b. Use hours worked versus cost; use hours worked versus charged.
   c. Account for material usage per job.
      - Include purchased materials.
      - Include rental equipment.
      - Include free issue tracking of small M&R items consumed by capital jobs and maintenance.
      - Set up contractors in the accounting system.
2. Determine the allocation basis for vendor-stocked material costs.
   a. Have a process for removing from stock.
   b. Have a process for accounting.
3. Manage M&R budget costs.
   a. Develop a process for tracking real-time M&R costs (rather than being surprised at the end of the month).
   b. Provide weekly updates to management.
   c. Interface daily with M&R management on spending status.
4. Track cost comparisons and trending.
   a. Fork truck and equipment repairs
   b. Small tools and materials
   c. Annual cost savings

   d. Comparison with other sites and corporate levels—benchmarking
5. Follow up vendor reports and bad charge corrections.
   a. Review maintenance, capital, and noncapital costs.
6. Develop an FTE capture process.
   a. Include company and contractor crafts.
   b. Include existing subcontractors not tracked by the hour.
7. Utilize rental equipment efficiently.
   a. Cover both maintenance and manufacturing.
   b. Have a process for tracking who is using.
   c. Have a process for determining when to release.
8. Develop capital and M&R budgets.
   a. Develop a process for project submittal to the budget.
   b. Obtain accounting cost distributions and work into the budget process.
   c. Develop manufacturing and site budgets.
9. Participate on the Accounting Team.
10. Track capital project costs.
    a. Maintain accurate project expenditures.
    b. Update the forecasts for year-end spending.
    c. Make cost accruals as needed.
11. Track the establishment, training, and turnover of all cost management processes.

Cost Control Process Team

If unsure of your present status or starting from scratch, consider forming a Cost Control Process Team to define the existing conditions (inputs and outputs along with the processes needed to connect them). This problem-solving process will capture the components of the "as-is" state. The SIPOC (supplier, input, processes, outputs, and customer) approach can be used to initiate discussions. Pick a pilot area to do this exercise to share learnings across the organization.

For a typical contractor maintenance organization, the following inputs can be documented to describe the things that we spend money on (costs flowing out):

Inputs

1. Mill supplies and consumables
2. Rentals (equipment, crane mats, portable toilets, etc.)

3. Materials (pipes, bolts, gaskets, etc.)
4. Subcontractor labor (NDE, etc.)
5. Affiliate support
6. Small tools
7. Office supplies
8. Breathing air (bottles, masks, etc.)
9. Welding supplies
10. Site labor (direct and indirect)
11. Employee recognition
12. Fit testing
13. Training (ISTC, skill, government required, etc.)
14. Equipment maintenance (tires, oil changes, PMs, etc.)
15. Fuel (vehicles, employee fuel cards, etc.)
16. Computers
17. PPE/safety (H_{2S} monitors, etc.)
18. Radios
19. Absenteeism (lost revenue)
20. Civil materials (gravel, sand, concrete, etc.)
21. Garage parts and repairs for sites (contractor-run shop with 3% markup)
22. Lawn equipment (may be available from another location or department)
23. Company-owned equipment (not on a time sheet)
24. Rework (charged to us to maintain a good relationship with the client)
25. Lump-sum lost materials
26. Late billing (write-offs)
27. Partial billing (get some versus none)
28. Out-of-scope work (disputes)
29. Overhead (reimbursable and nonreimbursable)
30. Contract negotiations (changes and add-ons)
31. Certifications (welding, etc.)
32. Hiring costs
33. Random (charged to job number) or post-incident drug testing
34. Delays and variances (hold up approvals for payment even when signed off in the field)
35. T/A field time sheets (document delays)
36. Paperwork
37. Out-of-plant justification

38. Central overhead charges (billing for client—not reimbursable)
39. Affiliate charges for equipment (additional to labor and sent in later)
40. Fleet manager costs
41. Management of client's subs (requested by the client but we have liability)
42. Insurance (backup coverage)
43. Injuries
44. Legal services
45. Overtime (OT) meals
46. OT payroll issues (disputes)
47. Gap between billing at client rates and higher wage rates for transfers to maintain core employees during slowdowns
48. Approved vendor list
49. Quality assurance/quality control

The following outputs describe the things that we do that generate money (revenue flowing in) or that cause delays in receiving funds.

Outputs

1. Labor
2. Company-owned equipment
3. Markups (rentals, subs, etc.)
4. Customer approvals
5. Time input to track
6. PO increases
7. SAP to track rejections (weekly resolution)
8. Caught mistakes
9. Start work without PO issued
10. Wrong PO number for work
11. Forced to charge to a different PO (first one ran out of money)
12. Client system blackout period (system changeovers)
13. Locked POs for processing (hold-up time)
14. Closed PO too soon (all charges not yet captured)
15. Equipment rate decisions (hourly, daily, monthly)
16. Contract addendums
17. Reputation (preferred provider for this area, good relationship)

18. Site relationships (all levels)
19. Competition support (shows our quality is better by comparison)
20. Additional specialty equipment and tools (not part of previous contract)
21. Availability of company-owned equipment versus renting with markup (need to know all available equipment for maintenance and affiliates)
22. Safety performance
23. Company transfers (supervisors on their tools and core workers paid their higher rates)
24. Company freight transport versus hotshots
25. Labor broker role (supply for garage, labs, embedded crews, etc.)
26. Zone-set head count (not backfilled every time—loss of hours)
27. Absenteeism (3% to 4% of total)
28. Capital flexing of head count
29. Layoffs and rehires
30. End-of-year slowdown (budget spent or cost-saving efforts by client)

The following processes were identified to link the inputs and outputs to describe how things happen for cost control (manage cost and revenue). Once the list of processes was completed, the three team members were each allowed to spend $10 to rank those processes that they felt had the most impact on cost control. They could spend a minimum of $2 and a maximum of $5 on any single item. The number shown by the process is the amount spent on that item based on the combined votes. As you can see, six of the processes were chosen.

Processes

1. Cost control (data flow, etc.), $7
2. Timekeeping (labor and equipment), $6
3. Estimating and planning, $6
4. Invoicing, $6
5. Aging and accruals (hard to do accurately), $3
6. Equipment tracking (catchall), $2
7. Procurement (order to receive)
8. Consumables and consignment (special tools, etc.)
9. Collections (old outstanding resolutions)
10. Central invoicing status versus site backup (tie together and match)

11. Client/company/central reconciliation
12. Systems optimization
13. Work execution
14. Quality control
15. Payroll and track (two systems—duplicate entries)
16. Posting of job costs (four processes now)
17. Hiring
18. Safety
19. Terminations
20. Employee movements (keep core workers busy)
21. Customer approvals
22. KPIs
23. Continual improvements
24. T/A and capital document management (different from maintenance)
25. Reporting and document management
26. Rework resolution
27. Engineering firm interactions (drawing errors, etc.)
28. Scheduling
29. Billing package issues
30. Procedures (maintain continuity, train new hires, etc.)
31. Vendor qualifications

The team can then develop the "as-is" process map (with details for how it happens now even if incomplete and fragmented) for the present cost control process. Combine with any support documents or practices that you are now using to help fill in the missing pieces. Start with the "big steps" to form the initial blocks in your work flow diagram. Once it is agreed that the as-is map is adequately detailed to understand, the team will create its "to-be" state to show the vision and identify the gaps the team will need to address to improve the processes for continual improvement.

To develop the input needed to create your "to-be" process map for cost management, the following typical notes capture a team's feedback for the required big blocks:

Cost Management Elements

1. **Labor.** Three systems need to be consolidated into one: time sheet, site payroll, and client system (required by the customer to match the gate logs).

   a. You must define the data needed for each system with the reasons and the value each step adds. Someone must take ownership to collect examples of each spreadsheet and system report or template to study and identify missing items, duplicate data, and wrong information. This will allow understanding of the required content for the single-entry template to capture the data needed by all participants to do their jobs.
   b. With such a proposal, you will have the details needed to sell any necessary changes to the client to get buy-in and support to go forward.
   c. It is essential to minimize rework by central and the field (labor PO issues, etc.) to ensure that the output is accurate and timely (closer to real time), that it is shared with all parties, and that it is visible from the system for use in analysis.
2. **Job costs.** Multiple spreadsheets need to be combined into a common system.
   a. Details for cost are provided from bids for vendor purchases.
   b. Central receives invoices from vendors and puts the invoices into the computer system. If an invoice gets sent to the site, the folks at the site e-mail it to central.
   c. Purchasing guidelines are needed when converting to POs when an order is placed (capture commitment at that time).
   d. Vendors are not giving price information initially, causing delays and resulting in the need to follow up. Count and quantity are on the packing slip but not cost.
3. **Equipment.** Similar to labor, three systems need to be consolidated into one.
   a. Equipment used by crews is not tracked for hourly use (bill by the day) and can be used on multiple jobs. Utilization needs to be tracked to the job level.
   b. You must make sure the captured cost matches billed to site for contract issues and for site job cost tracking.
   c. An asset management process is vital to know what equipment the company owns, where it is, and what is being paid for.
   d. Equipment is captured on time sheets for daily use on jobs and continual use applications. All vehicles are allocated to all jobs.
   e. Equipment must be matched to the people using it.

   f. A fleet database is needed for tracking and knowing availability without calling every time. You have to be able to forecast needs and make reservations for pieces of equipment.
4. **Variances.** Labor and equipment must be captured in real time to give the full picture.
   a. Why do you have the margins you have (desired versus actual)? You need to work to change them.
   b. You must improve how you capture change orders, delays, etc., that impact the accuracy and timing of billing.
   c. You need to know and document the scope of work to know if requests are outside the scope.
   d. You must improve your auditing system to track paid versus billed. Central creates a company invoice number for tracking on a spreadsheet, but it's hard to see which bill was paid and which is outstanding.
   e. Company-owned equipment that is not on a time sheet falls through the cracks because there is no invoice associated with it.
5. **Reports.** These must be more timely (close to real time) from the company, and they must be visible from a common database that allows access to needed information.
   a. You need actual labor versus estimated labor to the job level with feedback given to all parties.
   b. The cost by job needs to have details available from one source.
   c. In regard to consumable usage, you need to determine what is being used by person and by job, what is being bought in bulk, etc.
   d. Affiliates and small sites should set a facility target.
   e. You must know and use your contracts to your advantage to manage costs and revenue.
   f. Nonreimbursable and reimbursable, taxable and nontaxable, should be captured by order category and type of work. The proper job setup process will allow data to be captured right up front.
6. **Job creation.** This is the best way to ensure proper PO setup with known information.
   a. The system must be able to capture client POs with its company job numbers and allow for updates (increases from change orders).

b. People need to have access to reports and to be able to pull the needed information to manage their jobs.
c. The company site personnel will monitor and track the percentage of the total spent to tell the customers when they are close to the limit. You will forecast the new total to finish. PO-level tracking is required by clients for their budget control. The company must manage at the same level as your customers to allow audits to match.
d. You need to have contract-level outstanding payments with drill-down capability.
e. You must be able to pull reports by work type.

7. **Resolution process.** There need to be rules on when and how to escalate to higher levels to get the needed attention to clear up outstanding payments.
   a. A resolution process is needed to deal with misapplied payments.
   b. An individual resolution process is needed for payments over 60 or 90 days past due.
   c. The resolution process calls for regular meetings (Partnership Team) where both sides submit agenda items for discussion.
   d. A list of approvers, their backups, and upline mangers should be created for escalation of requests after a set period of time between each request.

Continue with the to-be process map for your cost management until it meets all your needs. Review the set metrics to track utilization for people, equipment, and services so you can properly manage them using your new process.

Create Mutually Beneficial Partnerships with Contractors and Vendors

All owner sites must have processes developed to efficiently manage the spending on external services (firm bid contractor or vendor jobs and materials) and on external labor (contractors paid by the hour including materials used). Treat this as your own checking account and be able to explain all contractor expenditures and to understand the hours charged. First, you have to get a handle on who is writing checks on your account. All site requests must flow through a centralized contractor coordinator to truly understand the needs and to match them with the best resources.

Processes to Manage External Services

1. Implement an invoice verification process that achieves the following:
   a. Create a check and balance system that captures hours worked and materials used for comparing with the monthly invoice charges.
   b. Have meetings with key contractors (on-site labor providers) and establish routine follow-up reviews.
      - Gain consensus on expectations with established performance measures.
      - Get estimates of hours to perform work to compare with actual charges and incorporate into an overtime management process.
      - Require an annual plan for improving the value and quality of the services to be provided.
   c. Ensure that monthly billing expectations are met.
      - Create an invoice summary charge sheet (listing all active purchase orders).
      - Itemize each PO with a spreadsheet of work orders charged (hours and materials). Account for site materials used through the store's system.
      - Match the daily time sheets with invoiced amounts and only pay hours that agree with the site log sheets from the guardhouse. It is the contractor's responsibility to explain and justify discrepancies to be paid the next month if approved.
      - Provide documentation for all third-party services.
        - Markup applied
        - Copies of subcontractor invoices
        - Copies of receipts for materials purchased and supplied
2. Capture regular and overtime hours to explain gaps with the actual head count. A full-time equivalent is equal to 2,080 hours per year. An example would be that a contractor has two people on site daily but the FTE for total hours worked is averaging four people, and so action would be required to understand why and close the gap on expectations.
3. Maintain the 50-50 balance between labor and materials by requiring vendor spending management for external services for fixed fees and for material purchases.
   a. Site requests need to flow through a centralized owner/planning group.

   b. An SAP cost report needs to be reviewed monthly by the Site Management Team to ensure accountability.
4. Bid jobs and projects from multiple sources to validate contractor selection and to reduce cost. Document expectation for these jobs and get approval and support from maintenance management prior to bidding. Minimizing expenditures through cost comparisons is a way of life that must be established. Spot checks will be made to verify that existing in-house contractors remain competitive.

It is critical to the success of your company that all functional groups be given the authority and be allowed the time to focus on cost control and management. Develop standardized processes and procedures for managing all assets and labor resources. Then take the right actions to meet site and corporate demands to truly utilize your assets, minimize the cost, and eliminate surprises and missed opportunities. The present economic conditions add a higher sense of urgency for the success of this effort.

Properly managing your costs is the needed medicine to cure maintenance insanity. The next chapter digs in deeper to really understand that your entire workforce must be treated the same when it comes to planning and scheduling and job prep. We are all working under the "dome" of our site. You have to account for and manage all work for all parties to keep things from falling through the cracks.

CHAPTER 10

Contractor Productivity Improvement Process

The maintenance insanity cycle of doing the same thing and expecting different results is really tested when you move from a time and material (T&M) contract to a unit rate (UR) contract. You now set how much you will pay for a defined job (20 labor-hours total), but the area customer still wants to have Bubba and his 3-man crew do the work (32 labor-hours). Someone loses money in this scenario.

The following success option helps both sides understand the changes needed to transform your present organization when moving from T&M to UR to perform properly for all parties involved including the contractors. Unit rate work sets a price for a given amount of work to be performed based on industrial standards and historical data. The thing people do not understand is that planning and scheduling is not essential for T&M work processes, but the moment you move to UR, you are required to do a very good job of planning and scheduling for all parties. With T&M, the operating area can set the crew size, makeup by name if desired, and have people sitting around waiting to respond. With UR, the operating area only gets to establish the scope of the work and set the priority. The execution force has to be able to control the crew makeup and size to fit the cost model. The other critical requirement for UR to succeed is adequate lead time to prepare for the planned jobs and then minimum disruption to the schedule. True emergency work must be handled outside the UR agreement and charged as T&M since planning and scheduling is not possible.

Let me explain unit rate with this yard work example. A husband meets with one of the crews that come through the neighborhood and offer to mow your yard. The yard is a little over an acre with lots of landscaping. The husband and contractor agree on $85 for mowing and weed-eating the edges. On Saturday, the husband gives his wife the money for the mowers and goes for a round of golf. The crew shows up with four guys and has the yard done in 45 minutes. Since they got through so quickly, the wife asks them to trim some limbs and weed some flower beds. When they finish, they come to get paid and the wife only gives them $85 since that was the agreed price the contractor quoted her husband. In industry, we have the same misconception when we agree to a defined scope of work at a set price but demand that we get our same six-man crew all day to be available to do extra jobs but only pay what was on the schedule. If a unit rate job pays eight labor-hours, you can assign one worker for eight hours or eight workers for one hour each, but you cannot send two workers for eight hours because it is what we used to do (maintenance insanity) on T&M jobs.

We found that contracts are developed and approved to initiate unit rate work processes without fully communicating the requirements and gaining understanding and buy-in by all parties involved. The following process evolved through many iterations to align all parties and to create mutually win-win partnership agreements that benefited each participant. This is focused on a multiskilled contractor supporting operations, but the concept fits any maintenance and operations partnership for any size plant facility.

Figure 10.1 is a one-page P&S flowchart that shows the essential elements for a successful P&S process. Notice that emergency work totally bypasses all planning and scheduling and also disrupts all scheduled jobs. Changing or expanding the scope of a set job is another common roadblock to effective planning.

A complete PowerPoint with training slides for a team exercise is provided at www.maintenanceinsanity.com for your use in determining the level of understanding your crew has for P&S.

The planning and scheduling framework to support the flowchart in Figure 10.1 includes the following steps:

- The customer for your services provides a four-week look ahead, with Week 1 at least 80% accurate by Tuesday for each work type. Week 4 is only 20% accurate to allow us to order long lead time materials.
- Crafts initiate walk-downs to identify bundling opportunities for feedback to the scheduler on Wednesday for Week 1 input.
- Wednesday review sessions must be held with all involved parties (routine maintenance, projects, major maintenance, and turnaround input between operations and all execution forces).
- The frozen schedule is set on Thursday for next week's execution.
- Break-ins are defined as any jobs performed that were not on Thursday's frozen schedule.
- Operations controls work, and maintenance/contractor services manages the resources to maximize the workforce with flexibility to complete requested jobs.

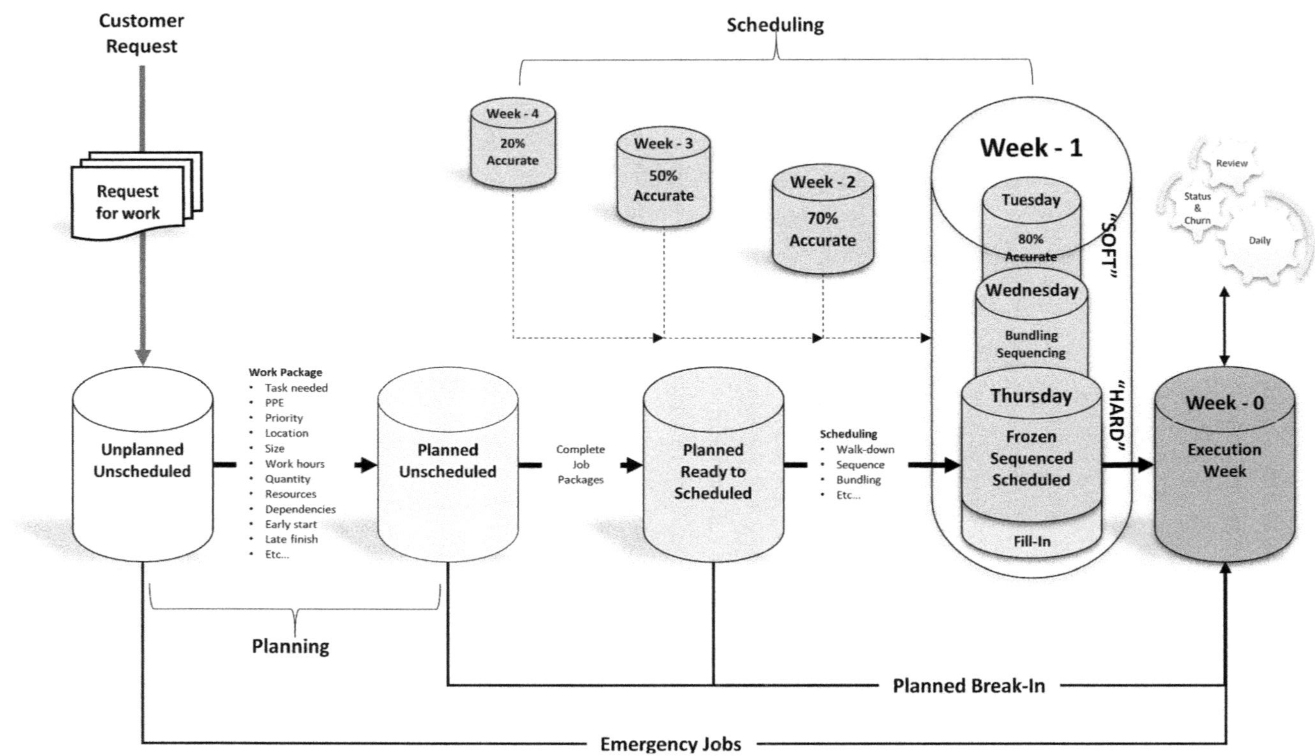

FIGURE 10.1 P&S flowchart

- Crafts finish work as scheduled to meet the agreed-upon completion dates. Fill-in preplanned jobs that require little support by operations are provided in case a scheduled job has to be delayed.

Explanation of the Productivity Coordinator Concept

The contractor productivity improvement process is for implementing identified contractor management best practices at any site. Start with one pilot area and then expand the process plant-wide. Start with one contractor and one operating area to fine-tune your process for all the services that they provide. Use your Thursday's frozen schedule performance to identify the positive impact for operations and maintenance. Expand your typical schedule to include the contractors for their weekly look-ahead projection of jobs.

Implementing the productivity coordinator (PC) role for the selected contractor will focus on field implementation coaching and efficiency enhancements for each area covered. The main value will result from properly planning and scheduling the work for the contractor's workforce, just as is done for the client's personnel. The PC becomes your field overview point of contact for all services needed. He or she will scope the work and get the appropriate person and craft involved at the level required to do the work most efficiently. Having the operations maintenance coordinator go through a single point greatly improves communications and results. Since typical contractors use their supervisors to do most of the detailed planning, this multicraft scoping role will be part of the productivity coordinator's duties to increase coverage and efficiency. This process will benefit all the services provided by all the contractors. The PC also becomes the crew advocate who listens to the concerns of the crew members and responds to support their needs. The PC takes a "big-picture" view to make the individual crews more productive by their interactions and job moves with minimum scheduled work disruptions.

The Core Team members from the pilot area, the contractor productivity coordinator, and the client lead should hold a two-hour kickoff session to explain the concept and gain buy-in by all participants. A sample presentation is available at www.maintenanceinsanity.com that you can modify to fit your specific needs. The job description for the new role of productivity coordinator is given below to allow you to understand the focus and expectations. If the selected contractor supports maintenance and operations, select one PC for the maintenance craft services and one for the operational support services to cover the entire plant. Have your contract services lead person facilitate the process and serve as coach for the process changes. This is an enhancement process to build off the already good planning and scheduling processes in place. The Core

Team will work together to eliminate roadblocks, identify gaps, and drive improvements to increase productivity for all the services provided by the selected contractor.

Contractor Productivity Coordinators: Responsibilities and Duties

The productivity coordinator will have complete responsibility for ensuring that all requirements and work activities performed by the field personnel are conducted safely and cost-effectively and with the proper quality to improve crew overall efficiency. This position will work closely with the contractor supervision, contract services, and maintenance coordinators in the field to reduce barriers that affect the productivity of craft personnel. The productivity coordinator is the number one controller of field productivity, ensuring that the weekly frozen schedule is adhered to and deviations are minimized. He or she will give input to the development of the customer's weekly sequenced schedule (including all contractor work) to help the crews bundle work and stage materials. Bundling input includes demos of scaffolding to use for new builds, grouping of hard and easy to achieve jobs, jobs in the same immediate area for material staging, and jobs to best fit the use of rack trucks or buggies. They help achieve area customer buy-in to gain acceptance for the proposed contractor input to the schedule job sequence. Ensure that accurate information is being shared with all owner and contractor crews. This addresses maintenance and capital work as well as operations support.

The productivity coordinator has the responsibility of coaching, mentoring, and training field personnel on productivity improvements and crew efficiency enhancements. This position will be an active member of the Contractor Improvement Team to help with field implementation of required data collection forms and to incorporate feedback from the field personnel to drive continual improvement.

The responsibilities of the productivity coordinator are to:

- Assist site and area contractor managers and supervision to help manage the processes related to productivity improvements.
- Audit enforcement of the weekly schedule given to the field team.
- Create forms and conduct periodic time-on-tools studies capturing delays in day-to-day work activities to drive efficiency improvements.
- Communicate with field personnel to ensure that they understand the purpose and instructions for the data being collected.
- Audit to make sure these forms are properly and accurately completed.

- Manage and maintain the productivity site action register: track and update existing action items, identify the addition of action items from field observations, and report the progress status weekly.
- Communicate with other productivity coordinators working together as a team to standardize data collection forms when applicable, share best practices, and develop methods for quantifying the savings related to improvements in productivity.

The productivity coordinator is a full-time, on-site position that requires approximately 75% of work-hours spent in the field interacting with the crews and the remaining 25% in the office. The productivity coordinator works through division management, and division management's approval and support must be obtained for the productivity coordinator to have the authority to do tasks defined in the PC's roles and responsibilities. Because this position interacts with all levels of management and field personnel, as well as all contractors and field coordinators, excellent oral and written communication and people skills are required.

Primary Duties

1. Manage the productivity action register including existing, ongoing, and new items. Facilitate the Pilot Leadership Team review of progress, exposure of barriers, and identification of opportunities to make the action register a "live" document to achieve continual improvement.
2. Work in alignment with the contractor site managers (execution feedback), admin (invoicing and time sheet input), and contract services (processes, systems, and field data collection forms), helping to mentor, train, and coach the field personnel on "how to do it right."
3. Actively work with all other site productivity facilitators to establish best practices to drive efficiencies. Align the site productivity action registers accordingly to drive all common best practices. Contribute to the weekly networking calls to benefit all sites with lessons learned.
4. Resolve productivity conflicts through face-to-face relationships whenever possible but escalate when appropriate to get additional support.
5. Conduct field input time-on-tools audits and observe the crews to collect data on what is affecting productivity. The information from these audits should be communicated appropriately to the crews, supervision, and man-

agement including the associated potential savings. Share the basis for the projected savings.

6. Implement forms in the field from contract services, contractor management, and administration for required data collection (billing, productivity, hours worked, delays, etc.). Conduct crew field audits and discuss the findings with the crew foremen.
7. Identify opportunities for productivity improvements and make constructive suggestions for change. Assign identified improvement projects to the appropriate cross-functional teams.
8. Select and train a backup person to cover your site-critical duties when you are not available.

At a minimum, productivity coordinators must

- Be subject-matter experts for crafts observed.
- Have the ability to manage conflict resolution.
- Have the ability to interface with various levels of management, customers, and operations/maintenance personnel for each zone, work unit, or area.
- Have excellent interpersonal, verbal, and written communication skills.
- Be a change manager to effectively introduce new ideas and drive to implementation.

The complete productivity coordinator checklists available at www.maintenanceinsanity.com will allow you to drive and track the required behavior changes and the desired results to be achieved by implementing a step-by-step process. This is a customizable form with three roles: the customer contractor lead, the craft productivity coordinator, and the operations support productivity coordinator.

The Core Team's primary focus will be to work on planning and scheduling activities, communication, and work flows to allow the selected contractor to be as productive as possible working at your pilot area. Key members of the Core Team will include the maintenance coordinator, operations area manager, maintenance team manager, maintenance planner/scheduler, and materials handling team manager (for noncraft services needed).

To show the value being gained by this pilot, we will need to capture and track the savings identified by the improvements made and the projects completed. The schedule performance metrics for this pilot will be the schedule compliance and break-in percentages to trend our improvement efforts.

For the pilot, work together as a true team to achieve the following goals:

1. Improve communications to get information earlier including field walk-downs to minimize the possibility of being late and always in a hurry to get the work completed.
2. Standardize the permitting process across all plants to eliminate long waits. Let the areas know who is coming so they can be prepared.
3. Close the loop on supplying needed information (drawings, specs, etc.) so you can deliver when operations wants work completed. Keep everyone in the loop on the status of each job.
4. Be more consistent from plant to plant on what is required to safely perform the work. This can vary from shift to shift in the same plant. This will minimize replanning for scoped work.
5. Standardize the P&S process so the same level of detail is provided and accepted by each plant. Require the same information and preparation for all jobs.
6. Close the gap between operations and contract services so that the contractor receives complete information on what is needed. This will enhance your P&S and your ability to make materials available when needed.
7. Share details documented with engineering and operations so that they know the contractor's needs and limitations and the order of execution. Division representatives on capital projects often do not keep operations in the loop on work to be performed.
8. Establish clear lines of communications with all groups requesting and performing work for the pilot area through better scheduling of all their activities.
9. Improve the feedback provided by each contractor when jobs are finished or obstacles occur.
10. Ensure that operations and the contractors set the correct priorities to prevent the miscommunication of needs and the resulting rework. Prevent second and third calls being required.
11. Simplify your work processes and shape them to make work easier to perform going forward.
12. Prepare the crews to "do things right the first time" through better planning and scheduling with the proper overlap coverage.

13. Reduce emergency and reactive workloads by better management of your resources and needs. Remind them that you will "be there" when needed for true emergencies.
14. Reduce delays and rework from misunderstood scopes and changes from original desires.

A schedule performance metric will be tracked to show the progress. The pilot contractor will share the frozen schedule for all known work requests as of 4 p.m. Thursday for all submitted jobs. This documented work list will then be compared with the actual work completed during that execution week. All work done that was not on the frozen schedule will be tracked as the percent break-in. Percent schedule compliance will be the work completed that was on the frozen schedule. Break-in plus schedule compliance must equal 100%. The lead for the process will collect and share metrics and achieved savings with operations, contractor services, plant maintenance division, contractor, and management to keep everyone in the loop going forward.

This is the standard agenda for the weekly schedule performance meeting on Thursday at 1 p.m. with all parties involved:

1. Review next week's proposed jobs by crafts to make any needed changes before "freezing" the schedule. It is critical to have the maintenance coordinators participate to get their input for known changes prior to Monday's start of jobs.
2. Discuss any fill-in jobs to make sure they are appropriate for the execution week. Fill-in jobs are planned work and can be done anytime without disruption. They offer a value-adding alternative when scheduled jobs have to be delayed.
3. Discuss last week's schedule performance including details for scheduled work not performed and break-ins to understand causes and propose corrections to prevent in the future.
   a. Capture the responsible parties and action items for follow-up.
4. Review the metrics and graph trends for actual performance up to this point.
   a. Check for percent schedule compliance, percent break-ins, and percent schedule adherence (# scheduled jobs completed/total # jobs scheduled).
   b. Post these graphs in the control room and contractor office area.

5. Include miscellaneous topics and learnings.
   a. Review any completed improvement projects.

For this process, the productivity coordinator is an equivalent role for the mechanics and contractors, just as the maintenance coordinator is for the operators. The MC is the single voice for operations. This PC role will be for all maintenance forces at small sites. The PC and MC need to have the authority to make changes and coordinate overall efforts with everyone following their lead. They are both responsible to do follow-up to make sure changes and improvements are implemented and shared. The following list describes the duties of the key players with which the PC and MC interact to achieve their goals.

P&S Roles and Responsibilities for Key Players

- **Planner.** Adds value for craftspeople through understanding and preparing job requests prior to the start of tool time (provides answers to crew questions)
- **Scheduler.** Assigns resources at the optimum time to allow the most efficient execution of each job request (resource coordination)
- **Productivity coordinator.** Scopes jobs and serves as a mentor for crews by providing field explanations for efficiency and productivity gains
- **Contractor services and maintenance.** Collect and analyze data to drive work process improvements.
- **Maintenance coordinator.** Provides a single voice for operations input, sets the overall priority, and arranges equipment preparation.

To keep the pilot going as we expand our efforts to other contractors and plant areas, it is critical that we all work together to identify and implement improvement projects as we encounter opportunities each day. Taking the time to capture the details allows the idea to be shared with other crews and sites. In the section that follows, I have set out initial criteria to help people understand what should be considered as an improvement project. Anyone can submit a project to be captured as part of the contractor productivity improvement process.

What Would Make a Good Improvement Project?

A good improvement project would be characterized by the following:

- It results in and captures savings (hard dollar and cost avoidance). It shows improvement to onstream or equipment reliability and has a positive impact on site profit.
- It utilizes a new innovative solution that demonstrates unique thinking or a different approach to solve a problem.
- It learns from other crews and takes advantage of a solution they developed (leverage knowledge from networking).
- The analysis and solution go beyond the requested work to completely eliminate a problem instead of just doing a quick fix.
- It creates or enhances a work method. The steps used are documented to ensure consistency and clear communication to the next craft worker to do the job. It simplifies the process we use. It is shared with other crews.
- The solution creates the opportunity for a reduction in labor-hours for the craftspeople to perform the work. The most efficient group to do the work is identified.
- It eliminates a pucker job or improves the safety of a major activity or daily job.
- The project decreases or eliminates a significant environmental impact for the site.
- The crew implements a "success" from another crew. The project results from benchmarking data that identifies a best practice from any source available.
- Technical support is used, and everyone's strengths and experiences are combined to eliminate roadblocks and solve the problem.
- It is data driven and has a positive impact on our measures.
- It eliminates job delays and captures the root cause to prevent a recurrence.
- It prevents a process or equipment failures (saves) or reduces the needed frequency for this job.
- It improves equipment maintainability or ease of operation of the equipment.
- The craft workers are able to enhance their skills as a result of the project.
- It results in material savings or reduces the time required to obtain the needed materials for the job.

Opportunities for Further Value Creation

To obtain maximum success, management will need to drive the implementation of this productivity improvement process for all divisions with rollout plans developed by a target date for a sense of urgency. Adopt the "under-the-dome" approach to P&S to include all the work being done in an area by anyone entering the dome (all client and contractor employees must be on the area schedule). Consider implementing a reinforcement plan with peer competition at each area of your site (or sites) to drive cost savings (dollars to the bottom line but requires an owner to push and celebrate success for crew buy-in). Make a big deal out of small achievements, and your maintenance insanity will be gone for good. The next chapter will tell you how to use these improvement projects as one part of an overall performance management plan to get everyone involved.

CHAPTER 11

Plug into Your Workforce

To really break the maintenance insanity cycle, you must get everyone involved in identifying what needs to change to make life at work better for all participants. Encourage your people to act like owners: What would you do differently if you owned this company?

Consider using this simple approach for your people and your contractors to facilitate crew meetings to have them answer the following six questions. These people are around your equipment every day and know many things that they could do to make improvements. Asking for their help plugs you into a valuable resource and allows them the opportunity to be heard. The productivity coordinator from the last chapter will become their spokesperson as he or she earns their trust.

Answering these six questions generated this brainstormed list of potential projects. This example contains some typical crew input for use with your teams to get them started. You can decide how much if any of these answers you want to share.

1. What improvements need to be made?
   a. Teach everyone to correctly operate and clean up the equipment.
   b. Verify the correct application of our equipment versus the service it is in and the products being handled.
   c. Have maintenance share knowledge with operators to transition the handoff of immediate-response care.
   d. Have maintenance share equipment knowledge with operators to enhance their skills and understanding of the impact of their actions on the equipment.
   e. Address vent system plugging issues.
   f. Increase the reliability of the secondary hot oil system.

   g. Evaluate the uses of steam and other plant utilities to minimize waste and improve the reliability of the steam boiler, air compressors, and other utility providers.
   h. Focus on housekeeping to enhance safety and plant appearance and to improve the functionality of the maintenance shop.
   i. Identify and correct failure causes for the process charging systems to reduce waste.
   j. Have operators work with mechanics to learn troubleshooting skills.
2. What long-term or repeat problems have not been solved?
   a. Your team can address this at a separate meeting.
3. How can we reduce planned downtime for maintenance?
   a. Have equipment ready at the appointed time.
   b. Have more advanced notice.
   c. Create better predictive and preventive programs.
   d. Create and implement failure reporting sheets.
   e. Keep PMs up to schedule.
   f. Continue to improve the permit process.
   g. Share what we know with operators.
   h. Learn from operators.
4. What activities do we need to stop doing to eliminate low-value or unnecessary work?
   a. We need to stop working on requests that do not have a work order (only E-jobs should be handled outside the planning and scheduling process).
   b. Repair quality is being negatively impacted by production dictating the time allowed to make repairs and to do PMs (pay me now or pay me later).
   c. Jobs like the recent freezing weather problems go better when we all get together and preplan our actions (do more of this).
   d. We must hold to agreed-upon shutdown schedules through better coordination between operations and maintenance (there is always a cost for changes and disruptions).
   e. We should develop a smooth transition plan to hand over operational activities from mechanics to operators (do it with them watching, have them do it with your helping, watch them do it, and coach and serve as a consultant).

   f. We need to ensure follow-through on the implementation of things we change. (Example: training CDs must be entered into the site training system for credit to be given.)
5. What would you change to make your job easier to do and more efficient?
   a. Get more cross-training.
   b. Get better equipment.
   c. Develop contingency plans for unplanned outages.
   d. Ask for training as needed.
   e. Share learnings from maintenance activities (in crew and throughout the site).
   f. Find mentors from other departments or sites.
   g. Make it easier for someone to get help while doing a job.
6. What can we do to make our customers happier?
   a. Your team can address this at a separate meeting.

Tap into the "Force" of All of Your Employees—Proven Performance Management Plan

To be successful in changing, you must first be dissatisfied with your present condition. If you are happy where you are or think that you are doing well enough, you will not be motivated to change. Sometimes people infected by maintenance insanity do not even know there is a better way. Make a plan and work it. Be data driven and pull the trigger. Never lose faith that you will succeed. Get approval for your plan and use projected savings to justify these activities and to gain management support. Set high expectations for your organization. Do not limit your people by what you think they can do. Challenge them and let them say when you are pushing too much. Celebrate small successes to build to big ones. Share the results with all employees to get their involvement. Remember that what gets rewarded is what gets done.

Your people are the key to your success. Focus on the human side for the most effective long-term changes to your culture. Each player must understand "why" you are changing and "what is in it for me." All reliability starts when your skilled craft worker installs, rebuilds, or repairs any piece of your equipment. Every mechanic has to know and believe that what he or she does makes a positive impact every day. Make sure that this "impact" is communicated to management and to the entire site. We each need to feel appreciated for the value-adding contributions that we make. Create performance management programs that implement a fun process to capture achievements, encourage reinforcement by all parties, provide feedback, and develop pride in the progress being made.

Fish! is a book by Stephen C. Lundin, Harry Paul, and John Christensen. You may remember, it was first mentioned in Chapter 3. And as mentioned in Chapter 3, the authors propose four concepts to be used to boost morale and improve results:

- Choose your attitude.
- Play—make work fun.
- Be there for your coworkers.
- Make someone's day.

The best way to incorporate their concepts into the crews' daily work activities is to link the concepts with improvement projects giving each person the opportunity to be recognized and thanked for his or her contributions. The following results demonstrate the power that can be gained by plugging into this available labor and knowledge resource.

The Gone Fishing reinforcement program was applied to 11 small industrial sites over a 3-year period with the following results. Our team improvement projects were started in midyear of the first year with 24 completed by the end of the year for $321,000 savings without a performance management program being used. With the implementation of the crew-level performance management programs, 166 projects were completed in the second year for $4.1 million, and 204 projects were completed for $3.7 million for the first 10 months of the third year with the Gone Fishing and NASCAR reinforcement programs. These reliability and productivity process improvements also had a positive impact on product quality and safety performance trends. With the Gone Fishing program, crews earned ounces to see how big their monthly fish would be, and with the NASCAR program they earned miles to see which car won the monthly race. Peer pressure and friendly competition add fun to winning the bragging rights.

The basic rules can be changed to fit the theme selected. For the mock Fishing Tournament, the boat rules specify that each site maintenance and contractor employee represent one angler in the boat. The members of each crew will pick a Boat Captain from their team to share monthly tally sheet scores with their supervisor for the crew activities to determine the weight of their monthly fish size! The captains compare fish stories and agree on the weight of each fish to be awarded to each of their crew teams. For networks of multiple sites, a monthly conference call with the site teams and designated Game Warden will finalize the results to be reported to all the captains. Recognition was given for the biggest fish monthly and for the overall boat tournament winner!

Create a crew improvement project form to capture information vital to keeping track of the project. The crew Boat Captain gets the supervisor to approve and sends the completed form to the Game Warden for feedback and recognition.

Here is an example of an improvement project form we used:

Site/Contractor Crew Improvement Projects

Date completed: ______________________________

Team members (list the names of the contributors): ______________________________

Project description: ______________________________

Key initiative area addressed: ______________________________

1. Customer satisfaction
2. Safety/environmental
3. Equipment reliability
4. Cost savings
5. Quality
6. Personnel development

What measure and work process does this affect?______________________________

Status and results achieved:______________________________

Basis and calculation of savings: ______________________________

Supervisor: ______________________________

Send the completed form to the Game Warden for feedback and recognition.

To make it more fun and to increase the recognition provided (all participants can make nominations), monthly awards were shared for the following categories:

- **Biggest Fish.** The project earning the most ounces for the month (project bragging rights for the month)
- **Biggest Stringer.** The crew with the most ounces for the month (overall monthly bragging rights)
- **Barely-a-Keeper.** The most unusual approach to solve a problem (might not feed the whole family but it sure tasted good)
- **Ugly Stick.** The best safety idea (did not land a fish but the log gave you a good fight)
- **Hot Bait.** The best crew innovation implemented and shared with others (helps everybody catch more)
- **Honey Hole.** The best application of the *Fish!* (you know where to catch the big ones, and you are willing to tell the rest of us)
- **Snagged Fish.** The most progress made on an identified project (you have not got the fish in the boat yet, but you are wearing it down)
- **Skunked.** Illegal fish or hooked by snagging (the Game Warden rejected it, and you have to let this one grow a little more)
- **You-the-Daddy.** Stirred the water the most to keep the tournament going (you may not have the fish in the boat, but you sure are fighting hard).

The key to the success of this type of program is to have an energetic owner driving the program. That person works with the Site Maintenance Leadership Team to determine the desired behaviors and results that you want to achieve. Once expectations are set, the owner interacts with the crew teams to finalize the monthly tally sheet (sample below) to use for scoring and to drive activities. Do not limit your people by what you think they can achieve. Let their results and feedback show what they can do when they feel appreciated for their contributions.

Gone Fishing Tally Sheet

Crew: ________________ Month: ___________________ Ounces: ______________

Customer Recognitions

A. Gave a verbal report to the crew supervisor (1 ounce)
B. Shared a written report with the crew supervisor (3 ounces)

C. Served as an example of making maintenance a site issue (5 ounces)
D. Helped solve, proactively prevent, or eliminate a customer complaint (5 ounces)
E. Improved communications between operations and maintenance (5 ounces)
H. Satisfactorily responded to a customer complaint with a solution that prevents it from happening again (5 ounces)

Documented Team Projects (includes crew team projects)

A. Each implemented project (10 ounces)
B. Total project savings for the month (1 ounce per $1,000)
C. Progress made for active projects (1 ounce per project)
D. Project # 25, 50, 75, and 100 in the running list (20 ounces each)

Work Method Improvements

A. Implemented solution from networking with another crew or site (5 ounces)
B. Identified a "cause" for lost production and improved reliability (10 ounces)
C. Held a monthly review with operations to address lost production reasons (10 ounces)
D. Created a PM plan for a proactive/predictive activity (1 ounce)
E. Made a permanent change to reduce cost or time to do a job (2 ounces)
F. Reduced planned downtime for maintenance (3 ounces)
G. Proactively identified a problem with a piece of equipment prior to failure (10 ounces)
H. Completed the transfer of tribal knowledge to standardize on a best method and to become like-minded (10 ounces)
I. Served as examples of site interactive support between operations, maintenance, and technical support (2 ounces)
J. Demonstrated the use of measure and ownership to drive improvement (5 ounces)

***Fish!* Successes (2 ounces for each application example)**

A. Chose your attitude
B. Made work fun—found a way to play
C. Was present for your coworkers or customer
D. Made someone's day
E. Has site award winners included on last month's award list (2 ounces)
F. Got someone else to try *Fish!* (10 ounces)
G. Spotted someone creatively extending our vision (10 ounces)

H. Gave constructive feedback in a positive manner (5 ounces)

Safety Attitude and Performance

A. An unsafe situation was eliminated (5 ounces)
B. The crew went the entire month without an injury (15 ounces)
C. All mechanics know the score (days without injury, 5 ounces)
D. A 100-day celebration was held for the crew (25 ounces for first, 50 for second, etc.)
E. A behavior checklist for prejob reviews was implemented (2 ounces each)
F. A safety crew huddle meeting was held at the beginning of the shift (1 ounce)
G. A cleaner way to do a job was found that improved housekeeping (5 ounces)
H. An ergonomic risk was reduced by changing a work method (5 ounces)

Training

A. Training modules were completed by mechanics (2 ounces each)
B. Maintenance awareness modules were completed by operators (2 ounces each)
C. Maintenance tasks were shared with or transferred to operations (5 ounces each)
D. Operations knowledge was shared with mechanics (2 ounces each)
E. OJT requirements for an identified job were successfully completed (10 ounces)
F. Site capability improved through a training session from a vendor or other source (2 ounces)
G. A training memo that solved a problem was rolled out and implemented (5 ounces)
H. The key function has a trained backup (5 ounces)

Results: Best Ever for a Month/New Record

A. Lowest maintenance-related lost production (10 ounces for each crew)
B. Lowest M&R costs (10 ounces for each crew)
C. Lowest overtime (5 ounces per crew)
D. Fewest callouts (5 ounces per crew)
E. Highest percentage of scheduled work completed (10 ounces)
F. Solved a top 10 high-cost contributor from accounting data for reliability measure (10 ounces—monthly identified "high cost" issue was resolved)
G. Zero overtime (25 ounces per crew)
H. Extended safety streak for x number of days without an injury (10 ounces per crew)

I. Best attendance (lowest number of occasions per employee per month) (10 ounces per crew)

Total Monthly Ounces: ______________________

The scores and progress are shown on a "stringer" to compare the size of the fish for each crew. Figure 11.1 shows examples of fish on the stringer that can be adjusted in size and appearance to represent the crew's catch and make the program fun. The stringer file is available at www.maintenanceinsanity.com for use with your program. Get creative in the fun and current themes you use to get your people excited about the process. For a crew that does not turn in any results, you can show a boot or fish skeleton. Awards are given each month for the Big Catch (heaviest stringer and single biggest fish), and many specialty prizes are given at the end of the tournament participation in addition to the overall Biggest Stringer and Big Fish for the year. Giving small prizes like two movie tickets will get family members involved when they ask where the tickets came from and how they were won. You are creating memories with stories to retell as you keep the momentum going for your next peer competition program.

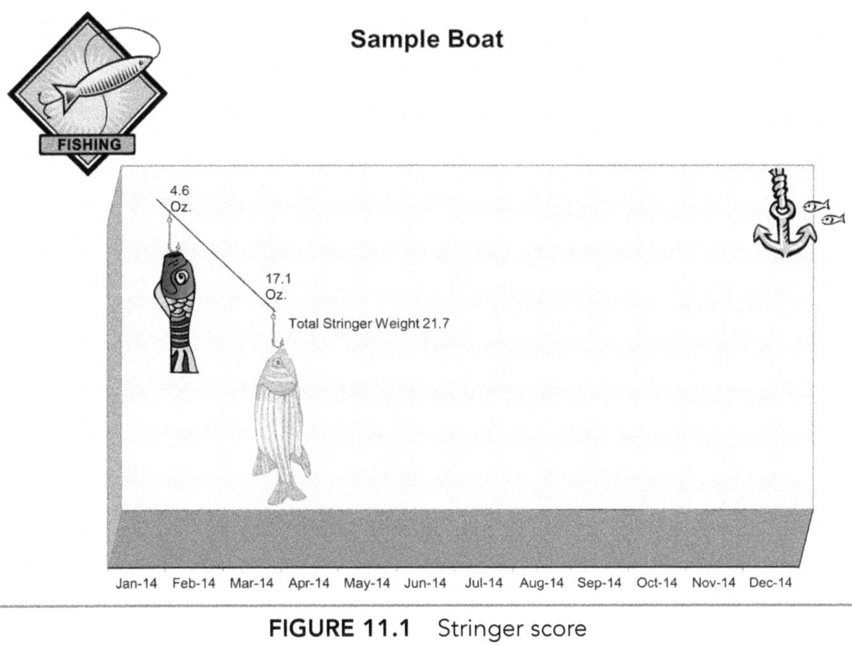

FIGURE 11.1 Stringer score

A monthly newsletter with results and winners is shared with participants and all other site personnel to get everyone involved. Here is an example of a *Fish Feedback Newsletter* we used:

The Game Warden threw back some illegal fish: Projects must be submitted on Project Form to be eligible for weigh-in.

Illegal Fish
No project form submitted for:

1. Using ASI-Net technology to install automated destination valves, the team deferred the cost of capital ($50,000)—Roebuck.
2. Installing load cell on the 250-gal Rx. ($2,500)—Roebuck.

December Weigh-In Results

Biggest bass was reeled in by Jefferson for C-9/Hydro filtration improvement weighing 510 pounds.

Team members: John K., Don, Jim D., Joe, Susan, Tom, Mike, John M., Ed, Jim T., and Dan (off-site)

Boat Stringer weight (ounces) status is shown below:

	Jan and Feb	Mar	Apr	May	June	July	Aug	Sept	Oct	Nov	Dec	Total Stringer Weight
1. Chicago	4.6	17.1	7.2	8.3	1.5	1.8	15.9	10.1	14.8	17.6	75.6	174.5
2. Atlanta	19.1	7.6	7.8	5.2	1.1	7.7	23.4	29.7	7.8	49.3	6.9	165.6
3. Southwest	26.3	6.6	2.4	20.9	10.2	5.7	5.4	2.1	31.1	4.6	6.3	121.6

Through the month of December, Gone Fishin' has documented 166 projects worth $4,066,017 in savings. Not a bad stringer of fish!!!!

You-the-Daddy

1. Marco J. (Chicago) for winning the 2002 Fitness Award
2. Rod G. (Atlanta) for multiple R+'s

3. Sanjay H. (SW) for developing a centrifugal pump guide for folks in the region and J.T. for passing it on to the sites
4. Sabra E. (SW) for helping correct charges for Moundville spare parts moved to Roebuck
5. Mark S. (SW) for his help with the start-up and debugging of the Latex Expansion project
6. Bruce L. (Atlanta) for helping with lots of requisitions so projects could be completed by the end of the year
7. Terry B., Bruce L., Ben S., Whitney S., David R., David L., Gregory L., Dwain F., James G., Ray H., Bill H., Gary O., Lewis S., Donnie S., Tommy S., Vanette S., Elaree S., and Gilbert D. (Atlanta) for completing one year of work without an OSHA recordable injury
8. Sam B. (Chicago) for lab excellence
9. Bryan J. and Scott T. (Chicago) for K-4 solvent recovery
10. Scott T. (Chicago) for scrubber improvements
11. Jon C. (Chicago) for mix tank spray balls
12. Jason F. (Chicago) for late ammonia addition and cons vent pluggage
13. J.T. (Chicago), Scott T., Tony N., and Campbell K. (SW) for K-43 head replacement
14. Joe W. (Chicago) and J.T. for thermal oxidizer liquid carry-over
15. Ron T. (SW) and Larry L. (Chicago) for slanted bottom on M-6
16. Jeff S. (Chicago) for safety control head replacements
17. Carlos M. (SW) for cross-connect hot oil heaters
18. Ron W., Steven D., Dean G., Lamont B., Mike B., Ray L., Steve N., Jason S., Thomas E., and Brian F. (Atlanta) for making maintenance a site issue—teamwork and good communications
19. Nancy T. and Greg C. (SW) for Lynwood stores support
20. Kenny S. (SW) for early-day SOPs for another site

Ugly Stick

1. Rick N. (Atlanta) for HCl
2. Chris R. (SW) for EAP SOP

(continues on next page)

Snagged Fish

1. Erick R. (Atlanta) for storage tank level and stick maintenance

Barely-a-Keeper

1. Li S. (Chicago) for process restroom drain line work plan

Another critical element for success is leadership. Someone must know where the organization needs to go to make sure you are on the right course. The Site Leadership Team members must fulfill this role since this team is made up of members representing each group leading the P&S processes. To keep the level of excitement going, these leaders must mentor and nurture their employees.

Coaching Role for the SLT Members

1. To repeat what we said earlier, make work fun. Your people will be more productive and cooperative when they look forward to coming to work. Treat them the way you want to be treated.
2. When interviewing contractors or new hires, look for and select people who have the attitude you want. Without the right attitude, it will be harder to teach them the skills they will need. It is your job to draw out the talents they already possess and to give them new ones.
3. You are the person who sets the tone for your people and who serves as a role model. If you do not do it, how can you expect them to?
4. You are changing many of their existing paradigms and beliefs about your site. The best way to succeed is through consistency, simplicity, and repetition. Lead by showing—as just noted, you are the role model. Only when you believe in people will they believe in you.
5. Your employees are your most important resource. Treat them with the respect that you would someone special, because they are special. Success is as much emotion as it is result. A sense of appreciation nurtures the growth and output of each employee. By implementing an abundance mentality, you want your employees to achieve their maximum capability and allow you to share in their successes.

6. The maintenance environment must prepare employees with the skills and freedom to do their job the way they think it should be done. They will do what you want them to do when they share your purpose and values. You create the vision, and they make it happen.
7. Encourage questions and focus on explaining "why" when implementing changes, and you will gain what is in their minds and their hearts. Respect is a two-way street. It must be earned as well as given. If you take care of your people, they will take care of you and your site.
8. One of our biggest competitive advantages lives within each of us. When you believe that you make a difference and share that value-adding feeling with all of those around you, the impact multiplies and the results are phenomenal. This is your challenge as you grow your employees.
9. You are working yourself out of a job. Someone must be able to do everything that you do in the long run for you to be able to be considered for a better position. Observe your people's abilities, attitudes, and leadership skills to identify potential advancement openings and relief coverage assignments. Give people the opportunities to exercise new skills and coach their performance to ensure a smooth transition when it is time to pass the job on to one of them. Doing the best you can with your present job is the best way to get a better job.
10. It is OK to not know. If you do not know the answer to someone's question, admit it and start working to find the solution. It is easier to trust someone who is human and makes mistakes than it is to trust someone who appears to always be perfect (which none of us are).

• • •

With all of us doing our best, maintenance insanity will not even be a big concern anymore. Everyone will be too busy suggesting new ideas and improvements to drive the changes that will ensure the optimum future for the company. The next chapter helps to establish the right team structure to link the leaders with the doers so we all succeed.

CHAPTER 12

Proper Team Structure

A proper team structure is required to maximize your efforts when implementing a cultural change to any organization. You are just feeding your maintenance insanity when you leave the same structure in place and the same people in the same roles forever. When you take a look at your present condition and decide that you want to be something better than you are now, you have to get your people willing to give it a try. This proposed team approach for a company with shop facilities supporting the field maintenance site operations will give you an idea of the buy-in and understanding that has to be gained at each level to proceed with rolling out a developed vision that can be communicated across your entire company.

We started by having the direct reports to the CEO meet on a biweekly basis as the Management Leadership Team to establish lines of communication and to establish roles and expectations for interactions and support of each function and the entire company. Once the MLT was functioning, the structure below was created to spread the culture change across all organizations. Pilot programs were used to get quick wins and to document the processes for roll out to the other similar functions.

Team Structure and Implementation Resources

1. Company Management Leadership Team (MLT)—direct reports to the CEO.
    a. Reviews and approves.
    b. Sets direction and policy.
2. Selected pilots from each major division (shops and sites). Start by focusing on initial work processes and cost control. These teams are to develop the packages needed to roll out to the other support shops and maintenance sites.

   a. Create and develop.
   b. Try the concept at their locations (test and refine).
3. Company network teams. Shops and maintenance teams are to represent all areas. Teams include the Shop Network Team (SNT) and Site Circuit Team (SCT).
   a. Communicate and lead implementation by working out the details of processes, learnings, and best practices to be implemented.
   b. Directly roll out implementation across the entire company for all approved items from the MLT
   c. Initial focus areas include
      - The partnership agreement (effectively work together across all boundaries) and measures
      - Market capability sheets were developed to show each group's services and skills for sharing and promoting services for all parties. We developed "cheat sheets" so everyone would know and understand what the others were able to do
      - Standard morning reports to share information and resources
      - Monthly scorecards to drive continual improvement by focusing on the right behaviors and results
      - The project management process and job books for tracking actual versus estimated
      - The implementation of pilot processes and reporting format for cost control
      - The rollout of standard company processes—to ensure that "you have what you need when you need it" for people, materials, equipment, and resources
        - Staffing
        - Asset management (start with inventories)
        - Purchasing
4. Field implementation specialists. Hands-on resources are to visit sites and shops to drive full implementation of approved processes and work practices.
   a. Corporate champions for each team.
   b. Work process mentor. People skills to teach field processes and verify understanding and compliance by supervision and craft personnel.

c. Management process mentor. People skills to teach computer and administrative processes and verify understanding and compliance by office and management personnel.
   - Select potential future leaders and use this assignment as a training exercise and observation period. These assignments would be for six months to one year.
   - Provide hands-on coaching and mentoring on a rotating basis where the coaches and mentors stay at a site for a couple weeks and then go to another. They return for shorter stays to shape and coach and to verify continual improvement while they were gone.

Management's role is to feed-forward a futuristic plan of where the company is going and how it is to get there and to make clear decisions in coordinating and directing employees' activities. Feeding forward provides the basis for control at the point of action and decision making for all parties involved. Management is to use feedback from all levels to measure the effectiveness of control during and after implementation. Information must clearly flow from the top down and the bottom up.

The top-level MLT is the source of the company vision that is to be shared across the organization to gain alignment and determine the path forward for each group. This leadership team improves communications and reviews and approves company policies and improvements for implementation across the company. Each shop or site then has to determine its present state compared with the desired "to-be" state to create its path-forward plan to get there. The network teams and pilots allow opportunities to prove the work process maps that will be spread to all the other locations. They are your test cases to try solutions prior to full rollout.

The network and circuit teams are the working groups that vet each proposed improvement from the MLT and determine which best fits each work type, location, or function. The following list shows the typical focus areas that they address:

Site Circuit Team and Shop Network Items

1. OSHA training records in personnel files (how training is done and test results documented)
2. Inventory processes and equipment tracking
   a. Charging to jobs
   b. Missed revenue
   c. Indirect charges if idle

3. Auditing of 360 safety program
4. Use of exit forms and performance evaluations
5. Management of people movements between sites
6. Job book use and feedback to estimating (joint field walks, budget tracking, actual versus estimated)
7. Subcontractor's COIs and MSAs when coming to the site
8. Rollout of new cost codes and purchasing module
9. Pilot site learnings
10. MLT decisions: partnership agreement and metrics, etc.
11. Reports run and distributed from the accounting database and pivot tables
12. Staffing checklist
13. Central services checklists (AP, AR, billing, payroll, etc.)
14. Proper use of approved rental vendors
15. Company branding—share the vision
16. Indirect charges
17. Auditing of MLT rules and expectations
18. Best practices
19. Improvement project process
20. Services provided and being pursued
21. Knowledge of company expertise sheets for all shops
22. Training needs
23. E-mail and calendar enhancements
24. Crew productivity
25. Proper use of field forms and data collection (delays, change orders, etc.)
26. Release notices for the PO's tracking tool for estimating and sites
27. Planning and scheduling of practices and exercise
28. Flow of labor and cost reporting
29. Leadership skills
30. Folder setup and scanning of documents
31. POs/invoice checklist
32. Shop and competitor spend and action plans to address
33. Shop and people productivity
34. Resource utilization at sites and across sites
35. Monthly scorecard use
36. Request for quotes (RFQs) and communication process
37. 24/7 response and standard Monday-through-Friday staff coverage

38. Work processes and leveraging of available software
39. Meeting efficiency
40. Bid development process (win-loss, estimated versus actual, historical tracking, etc.)

The field implementation specialists come into play when the company is ready to accelerate the culture change and build momentum for continual improvement. The specialists start with the company vision and engage each assigned area one at a time.

To lead any group, you have to know where you want people to go to be able to get them there. The specialists start by doing the following activities:

- Observe present practices, work processes, and people's roles. Do this at your assigned site or shop as you learn the processes developed by the pilot locations.
- Talk to people at every level of the organization. (What is the present culture, and what do they know about the company's improvement processes?)
- Explain your role and what you will help people achieve. (What is in it for them? Why should they change?) Use developed checklists of expectations for roles and responsibilities and desired activities to track and drive change.
- Standardize where possible and share best practices. Most companies have grown so fast that standardization and the sharing of best practices are critical to their success moving forward (share the company vision).
- Work together with your team leader to improve the site or shop you are supporting.

Use people's existing roles and what you desire to be different going forward to get them to try the new tasks by showing them how it will make their job easier. The network teams for shops and sites are your sounding board and conflict resolution aids. The Implementation Team leads will run interference for their specialists. Small sites may need central support, and some people may have to perform more than one role.

Proposed improvements must be understood before people will be willing to do them. Maintain action registers and track to resolution all assignments. Introduce an improvement project form to document people's ideas and to give them recognition for their contributions.

The specialists will work in pairs to address administrative tracking systems and work process rollout. Your job is to "show by doing." Implement checklists from central

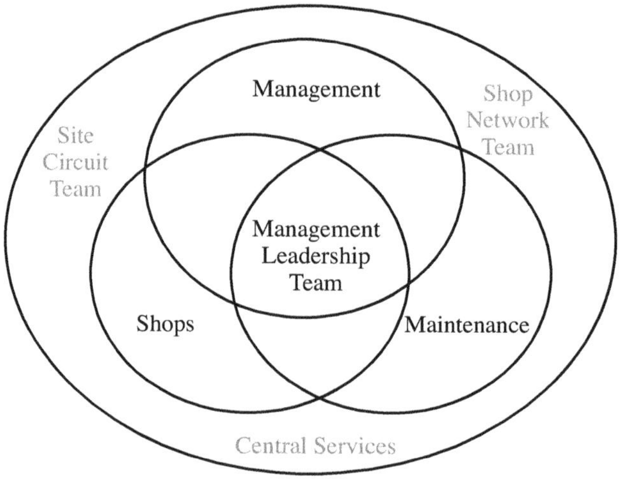

FIGURE 12.1 Overall team structure

services, staffing, and P&S to ensure compliance with cost systems and work processes. Coach the use of job books from your estimating group for all projects (standard form use and compliance). Track implementation status through a matrix score and send updates to implementation leaders of both Network Teams and the MLT. To free up the temporarily assigned implementation specialists, trained backups must be provided. This offers another cross-training opportunity for talented employees you want to grow.

The implementation specialists are trained on the company vision and expectations before starting their assignments in the field. They are given additional people management skills and directions for their field interactions. They start with the pilot locations their first week, and their leader then decides when they are ready to move to their second target to drive the identified improvements required to achieve the company vision.

Figure 12.1 presents the overall team structure for the company alignment. It shows the role of each team in changing the overall culture "under the dome" of the company. The specialists are members of the Shop Network and Site Circuit Teams but also leaders on their assigned Implementation Teams to provide hands-on coaching to achieve the change. They are observers as well as auditors, but with the twist that they get to help fix whatever they see is wrong.

In Figure 12.2 the site/company structure and vision are shared. The two implementation specialists work within this structure to support across and up and down the organization as described below. The management process facilitator (specialist) goes horizontally to tie the P&S to the business functions to make sure our people are given

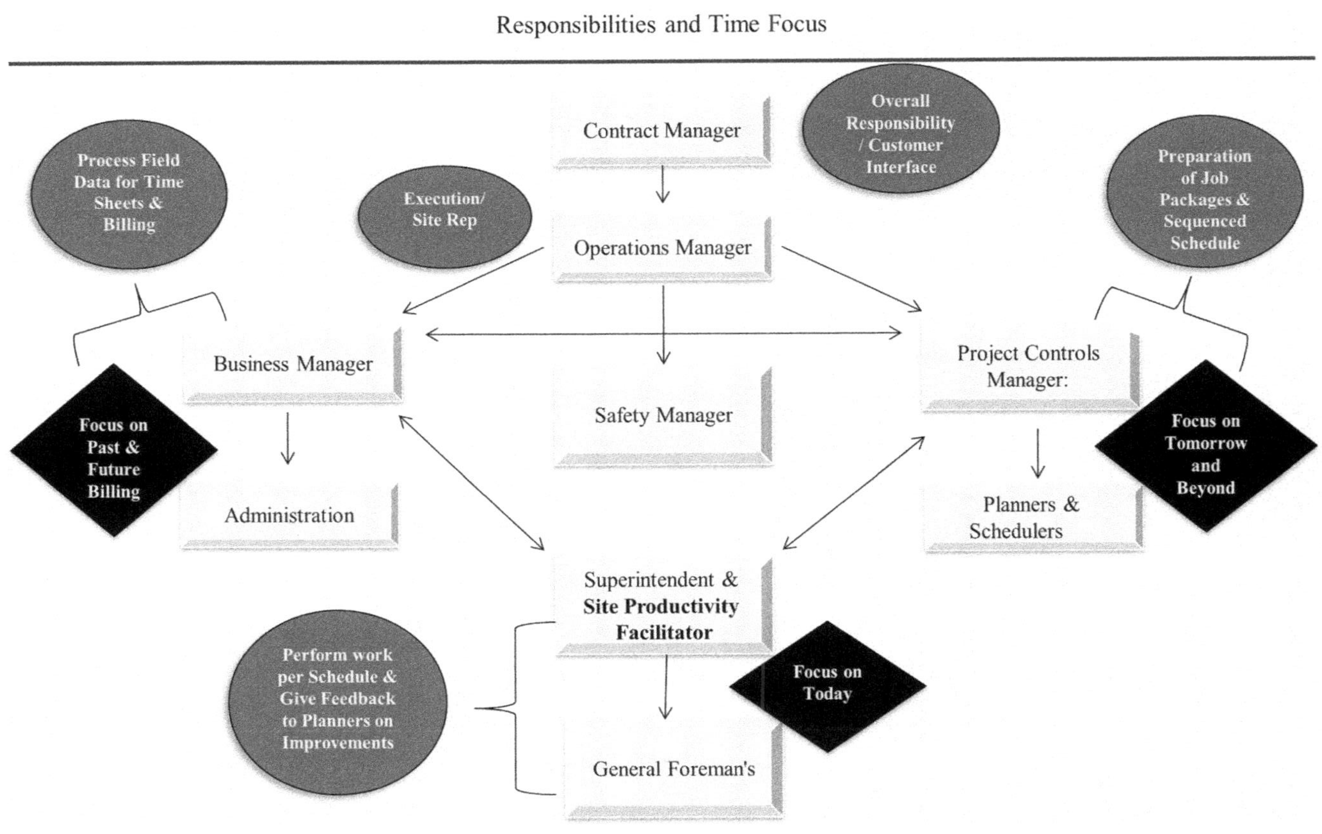

FIGURE 12.2 Site/company structure and vision

the information they need to do the work and we are collecting the proper data to get paid. The work process mentor (specialist) is focused vertically on getting the work done in the most efficient manner. These two specialists work together to ensure the needed data and forms are collected by the field so billing can accurately occur. Their Implementation Team leader must be in the loop and directing their overall direction.

The specialists will randomly use a customer feedback survey to score their impact on the people that they are serving. It will be good feedback to gauge their effectiveness and to identify opportunities to improve. One of the key opportunities for our teams is to start and spread the use of the improvement project process. Whenever you observe or participate in an improvement activity, make sure one of these project forms is completed and sent to your leader for tracking and recognition. This summary tracking captures savings to help justify the resources used for the implementation process. Use your reinforcement skills to make all the contributors feel appreciated. To succeed, you must get the hourly and salary employees all engaged and actively participating. Use one-on-one interactions and meetings to help solve problems and be more effective working with people. The specialists also perform training and problem resolution with the sites they support.

Have a goal to document "wins for the week" in each of your updates and during team meetings. The crew improvement projects make good examples for this topic. The team leaders must stay a week ahead of their specialists for scheduling their next site/shop visits and making sure they are ready for you to come. Copy everyone involved on all e-mails you share with the people you are helping. Always support and encourage each other to keep the momentum going.

Following is an example of an actual completed project by an implementation team. You can see how conservative the estimated savings are. It will be very important going forward for you to capture these projects and get the hourly employee and staff actively involved in the overall improvement changes. These will justify your efforts and allow management to expand your roles. It is really worth the effort and will be rewarding to everyone involved.

Site/Contractor Crew Improvement Projects

Date completed: January 2017

Team members (list the names of the contributors): Chris, Marco, and Teddy

Project description: Shop began tracking rework on projects and was transparent on where we currently stand and how we would like to improve. The team used the daily 2 p.m. production meeting to discuss rework and how it could be improved.

Key initiative area addressed:

1. Customer satisfaction: Reduced errors leaving the shop
2. Safety/environmental
3. Equipment reliability
4. Cost savings: Impact on final billing due to rework costs
5. Quality: Reduce rework percentage
6. Personnel development

What measure and work process does this affect? Rework

Status and results achieved: Shop's rework percentage went from approximately 27% in 2016 to 3.4% at the start of 2017. If the trend would have remained the same for 2017, with a 27% rework rate, the shop would have had 806 more red flags or reworked spools. By letting everyone know our failures and expectations, we were able to focus on reducing this number. It became a priority to all involved. The previous thinking at their shop was to get it done as fast as possible and we would fix things later. The new thinking was that if we slowed down and completed the work correctly the first time, it would speed up production, as we would not be doing the work two and three times. The annual savings is $40,300.

Basis and calculation of savings: Some of the rework is very simple, such as a small weld pickup or the spool piece missing a tag. To be conservative, we used a cost of $50 associated with each rework. When we multiply that by the 806 fewer reworks, that is a savings of $40,300.

With the right support and team structure in place, you can move forward by breaking the hold of the maintenance insanity and banish it by your new vision. Use the tools provided for you at www.maintenanceinsanity.com to strengthen your teams' performance. The next chapter will share a method to achieve continual alignment improvement to ensure all parties are operating on the same page and moving forward toward mutually beneficial goals.

CHAPTER 13

Continual Alignment Improvement Process

When we do not listen to each other, it is easy to stay bogged down in the maintenance insanity routine of doing the same things over and over and never really getting what we truly want—we just get more of the same. The continual alignment improvement process (CAIP) turns the client and service provider perceptions into a true situational analysis that leads to gap closure from the dialogue and action plans created by a selected Site CAIP Leadership Team with key players from each company. Cross-functional interviews capture perceptions of present conditions and cultural beliefs as your starting point. These findings are then compared with a set of benchmarked norms in a defined tier matrix to score the client and service provider input and perceptions for gap comparisons. A sample scored tier matrix is provided at www.maintenanceinsanity.com for your use to score your own progress as you implement the improvements described in this book.

The objective of the CAIP charter is to follow a defined process to identify, develop, and effectively implement continual improvement plans. The primary focus areas include safety, project controls, administration, execution, metrics, communications, and relationships and trust. The Site CAIP Leadership Team then develops alignment strategies with set objectives based on identified gaps in desired performance. The identification and sharing of best practices will be an important element as the teams move through the process.

Active participation is required from all levels and groups from maintenance, turnarounds, capital, programs, operations, and administration to accomplish the goals that are set and to keep the process and initiatives moving forward while providing improvements in either the process or cost savings. The client and the service provider will work as a cohesive team in alignment to maximize the performance of both parties.

Together they will successfully identify and track performance metrics to drive their continual improvement. A specific scorecard will be created to track and drive these target improvements. To ensure sustainable culture and behavior changes, the CAIP assessment is repeated on a set time schedule to verify gap closure. This process is critical for identifying best practices to be shared across a single site or multiple sites.

The CAIP process is a strategic advantage delivered through mutual teamwork to build a true partnership between the client and the service provider for a single site or a circuit of sites or regions.

To maximize the gain from CAIP, a network of teams is needed to support the overall process and ensure buy-in at all level of the organization. A Sponsors Team will initially meet monthly (quarterly long term) to be responsible for oversight of the relationship, provide direction and steward the progress of the Leadership Team, and serve as final issue resolution as required. The Leadership Team will meet biweekly (evolving to monthly) to identify initiatives and issues for overall improvement, evaluate resource requirements and delegate responsibilities to subteams for resolution, use a working relationship to steward the progress of tactical work and action items, share best practices, and monetize and capture benefits. Subteams will be formed as required to further the progress of initiatives and meet the timelines and goals defined by the Leadership Team.

Continual Improvement Tiers

In the list that follows, the benchmarked categories and levels have been refined over years of industrial experience and client interactions to serve as descriptions to help you identify your starting condition and then track your progress over time. The list can be used as your metric tool for your own performance or as a comparison tool to gain consensus between a client and a service provider. Level 1 covers the basics, and Level 5 describes best-in-class performance. Once your baseline is set, then you can decide what score to shoot for in each category and overall. A CAIP table is provided at www.maintenanceinsanity.com to allow you to mark your baseline and then periodically rescore to see your progress for each tier.

There are seven continual improvement tiers, and each consists of five levels.

Safety and Participation

Level 1

- The site safety plan is in place and is initiating positive cultural change.
- Safety is driven by regulation, management directives, and the cost of accidents.
- A proper focus and safety culture is defined.
- Basic result measures are in place.

Level 2

- Mutually agreed-upon programs are in place and functioning.
- Safety goals are being met, and employee participation is increasing.
- Safety has improved.
- Root cause analysis and prevention efforts occur.
- Safety measures show a positive trend.

Level 3

- Behavioral based involving all levels of the organization with safety culture institutionalized. Safety program is behavioral based involving all levels of the organization and the safety culture has been institutionalized.
- There is equal concern for people and compliance and injury costs, but they are seen as separate subjects.
- Safety is seen as equal to other functions, and a vision is in place to implement.
- A long-term process is initiated.
- Safety benefits are often discussed.
- Safety behaviors are captured in site measures.

Level 4

- Concern for people is a major safety driver, but this is not in harmony with other goals.
- A plan is in place to totally implement the safety process. Safety is becoming part of the overall process.
- Critical behaviors and conditions are measured. Long-term commitment is evident.
- The selling of safety is consistent and integrated. Employee involvement is "built in."
- All safety measures are trending positively.

Level 5

- A sincere concern for people drives safety and is in perfect harmony with other goals.

- Safety is totally accepted as an essential part of business and is an integral part of each employee's daily activities.
- Everyone knows that safety excellence is a never-ending process. An investigation is launched for every injury or near miss. Causes are identified and eliminated. Measures with goals are reviewed frequently.
- Safety benefits are an accepted part of the culture. Safety discussions are a routine part of everyday conversation.
- Measures are benchmark performance.

Project Controls (Planning and Scheduling and Costs)

Level 1

- Work decisions are reactive.
- Work is assigned as previous work is completed.
- There is no organized method for keeping track of the work needing to be done.
- There is no schedule to commit to.
- Not every job has a cost estimate (overkill).

Level 2

- Reactive work is reduced.
- Work assignments are given at the beginning of the week, but labor-hour estimates have not been done.
- Some scheduling is performed.
- A process is in place for providing cost estimates on planned jobs.
- Basic measures of performance exist.
- The customer is in control of backlog and issued schedules.

Level 3

- Key work processes are developed for proper function management and execution.
- A reporting structure is in place with deadlines established.
- Most planned jobs have cost estimates provided but no feedback to update the plans.

- Joint scheduling meetings are driving continual improvement.
- Work barriers are being captured.
- Reactive work is less than 50%
- Backlog aggregation has started.
- Crews are assigned to areas for ownership.
- Schedule compliance does not include churn during the week.
- Adequate lead time is provided to allow planned job package development for all scheduled jobs.

Level 4

- Uniform reporting procedures are developed for all crafts.
- Cost estimates are provided; jobs plans are updated as appropriate based on actual costs.
- Work processes are functional with continual improvement.
- Best practices are being shared.
- The backlog is mainly proactive.
- Effective aggregation is occurring.
- Managed programs have been implemented.
- The service provider controls the ready-to-schedule backlog and provides a weekly schedule.
- Performance measures are trending positively.
- Barriers are being eliminated as they are identified.
- A planning and scheduling protocol is established to benefit both parties.
- Schedule compliance captures all break-ins to drive accountability.
- Crew size and mix are managed.
- Work is bundled by areas including scaffold demos to use for new builds.
- Processes are adapted to best fit small, medium, and large sites.

Level 5

- Steps taken are proactive.
- Crews are consistently assigned a challenging day's work.
- Flexible "fill-in" jobs are identified ahead of time in case of delays or early completions.
- Crew size is the minimum required to safely and efficiently do the work.

- Schedule breakers are identified, and efforts are made by both operations and maintenance to eliminate their causes.
- The service provider gets customer approval of the weekly and daily sequenced schedules based on the provider's ready-to-schedule job packages and bundled work.
- Schedule compliance is measured against the sequenced activities to capture churn.
- Multicrafts are utilized to drive productivity improvements.
- Managed programs are adding value and providing a source for backlog work.

Administration and Business Decision Processes

Level 1

- The organizational chart is established, and partnership contacts are functioning.
- Reporting cycles are established with content defined.

Level 2

- Procedures are in place and are updated only if someone identifies a problem.
- Teamwork is shown in problem resolutions.
- Indirect costs are reduced.
- Basic financial measures are established.

Level 3

- Continual improvement is shown by periodic reviews with gap closure actions for all key administrative processes.
- All parties understand their roles and add value to the processes.
- Direct head count is being reduced.
- Monthly KPIs are driving improvement efforts.
- Site project control and the business manager structure are implemented by the service provider to manage the account.

- Measures keep process performance visible for driving the timely completion of identified issues.

Level 4

- Best practices are leveraged across business units.
- Business review meetings are held on a routine schedule to address ongoing issues.
- Day's sales are unbilled, and receivables are managed properly.
- Accuracy is maintained at a high and steady condition.

Level 5

- Best practices and all learnings are leveraged across the system.
- The Networking Team shares common processes with continual improvement tracked.
- Savings are captured to show valued added by the partnership.

Execution and Field Performance (Productivity)

Level 1

- Most required assets are in place.
- There is little cross-skill training for work being done.
- Reactionary work is still common.
- A single craft approach is used.

Level 2

- Factors have been identified that streamline job processes.
- A few of the people have skills outside their specialty, and schedule input tries to use them.
- Training is defined to initiate multiskills.
- Jobs are being bundled to leverage crafts
- Tracking of certifications is being established.
- All required equipment is available when needed.

Level 3

- The focus is on developing multicraft employees.
- Performance results show continual improvement.
- Most people are trained in more than one skill or craft and work freely in either area as assigned in the work schedule.
- Multicraft execution is under way per scheduled jobs that fit the criteria.
- Outage crews are shared to load schedules with resources.
- Productivity enhancements are tracked to change work processes.
- Productivity measures are tracked by craft and major jobs.
- Craft certifications are defined and are driving crew skill development.
- Planners use craft experts for input to special requests.
- Crews prepare the materials for tomorrow before they go home today.
- Fill-in jobs are identified to keep the crews productive when scheduled work is disrupted.

Level 4

- There is a high percentage of multicraft employees.
- Multisite resources are shared where appropriate.
- On-call productivity enhanced by multicraft workers allows a single crew to respond.
- Material management is optimized to prevent crew delays.
- Crew ownership is established by site area assignments.
- Planners use service provider experts for unique situations to leverage knowledge and enhance productivity.
- The schedule shows crew productivity targets for most jobs.
- The fill-in jobs shown on the schedule come from the two-week backlog of ready-to-schedule jobs.

Level 5

- Resources are leveraged across the system.
- Procurement is leveraged across the system.
- Crews provide feedback to improve stored plans and propose improvements for repeat work.

- Material is staged or demos are scheduled to provide the material for new scaffolds.
- Equipment utilization is optimized.
- Crew productivity targets are provided and tracked for all assigned work to compare with standards and estimates.

Measures/KPIs and Asset Utilization (Stewardship)

Level 1

- Little or no metrics or measurements are in place.
- No or an inconsistent review of performance exists.
- No goals or improvement strategies exist.
- There is a focus on "doing" as opposed to improving.

Level 2

- Some measures may exist in determining performance.
- Some review of performance exists. The review is typically limited to periods of dissatisfaction with performance.
- Goals or improvement strategies are available but are not being used to drive improvement.
- Crews respond to a variety of requests for data from both parties.

Level 3

- A template has been identified to set the baseline and tracking in place to share results.
- A quarterly review process is in place and is being used to identify efficiency improvements.
- All parties understand their roles for collection and analysis.
- Goals and improvement strategies are available for measures.
- Partnership meetings are under way to ensure mutually beneficial improvement projects.

Level 4

- Measures are used to make key decisions, serve as a motivator for innovation, and provide for evidence of the impact of improvements.
- Analysis flags opportunities.
- Results obtained are site specific.
- The total installed cost is reduced.
- Annual initiatives are identified.
- Customer management and service provider management are involved in the review process.
- Financial value is proved through innovation and proactive improvements.
- A scorecard approach allows site-to-site comparisons for sharing best practices.

Level 5

- Efficiency improvements are applied with results shared with other business units.
- The results obtained are mutually beneficial to the partnership.
- Advanced metrics include creating and applying relevant knowledge.
- Improvements have reduced the total installed cost for our systems
- The service provider shares learnings from other clients to enhance this partnership.

Communications and Expectations (Mutually Beneficial)

Level 1

- Limited expectations and guidelines have been communicated.
- There is no regularly scheduled review of performance.
- Communications revolve around "us" versus "them."
- Miscommunication is still causing issues.

Level 2

- Communications are shared with each other on a need-to-know basis only.
- Mostly verbal but some written communications occur.

- Processes are not linked to prevent duplicate efforts and miscommunication.
- Positive reinforcements are at a minimum.

Level 3

- The customer and the service provider participate in regularly scheduled meetings.
- The standardization of processes is under way with direct contacts being made.
- An ongoing feedback loop is initiated to keep all parties informed.
- Regular business meetings are driving the continual improvement of processes.
- Learnings are identified and shared to drive process improvements.
- Common goals are established to focus both parties.

Level 4

- There are direct interactions with customer business units including the proactive determination of expectations and needs.
- Appropriate contact is made at all levels.
- Knowledge transfer between sites is commonplace.
- Standardization is driven through shared best practices.
- Successes and learnings are shared and reinforced to drive the desired behaviors.
- Planning and scheduling protocols are incorporated into the short- and long-term sharing of desired work requests.

Level 5

- The customer and the service provider have a shared vision creating the greatest long-term value.
- Best practices are being shared and implemented across business units.
- Expectations are adjusted to keep moving to the next level of performance as improvements are implemented.
- Celebrations are shared at all levels for achievements.

Relationship and Trust for Customer Satisfaction

Level 1

- Limited relationships exist.
- The management of function is primarily performed by the customer.
- The service provider responds to all requests without consideration of the impact on overall performance.

Level 2

- Key roles are defined with customer and service provider counterparts established.
- There is limited recognition of mutually beneficial goals
- Most interactions are for problem resolution.

Level 3

- The service provider is accountable for function performance and improvement.
- Independent goals are set, recognizing the gaps between the customer and the service provider.
- Relationships are focused on mutually beneficial improvement items.
- Individual and team partnerships are growing.
- Working relationships are taking place at all levels.

Level 4

- The right people are managing shared interest with accountability.
- Touchpoints are aligned with business objectives.
- There is a strategic source provider.
- Collaborative gap resolution is the established practice to achieve win-win opportunities.

Level 5

- There is customer and service provider unification of all goals and initiatives.

- There is agreement that rewards are earned according to the value created for the organizations involved.
- A strategic partnership is established across the system.
- Safety, quality, and productivity are linked for optimum performance.

Figure 13.1 shows the overall CAIP scores, and Figure 13.2 displays the CAIP tier scores. These graphs show the scores for a comparison between a client and a contractor with the scores based on their perspective of performance levels. Cross-sectional and functional groups were interviewed, and then their input was used to score their tiers for each company. These comparisons were then used to create gap closure plans.

CAIP Steps

By understanding the tier ranges and categories for comparison, we can perform the established steps to determine our baseline and initial target areas. These identified gaps will lead our teams to make the needed improvements to achieve our long-term vision that will cure our maintenance insanity.

Checklist of CAIP Steps

Step 1: Tier assessment

Where are we now?

- ❑ Is the service provider's perspective of objectives currently in place and functioning properly?

What tier does the service provider maintain we are currently achieving?

- ❑ Is the customer's perspective of objectives currently in place and functioning properly?

What tier does the customer maintain we are currently achieving?

Step 2: Alignments and strategy

Alignment of the customer's and service provider's perspectives

- ❑ What is the gap in our perspectives?

Development of the customer's needs and the service provider's value proposition

- ❑ What tier does the customer want to achieve?

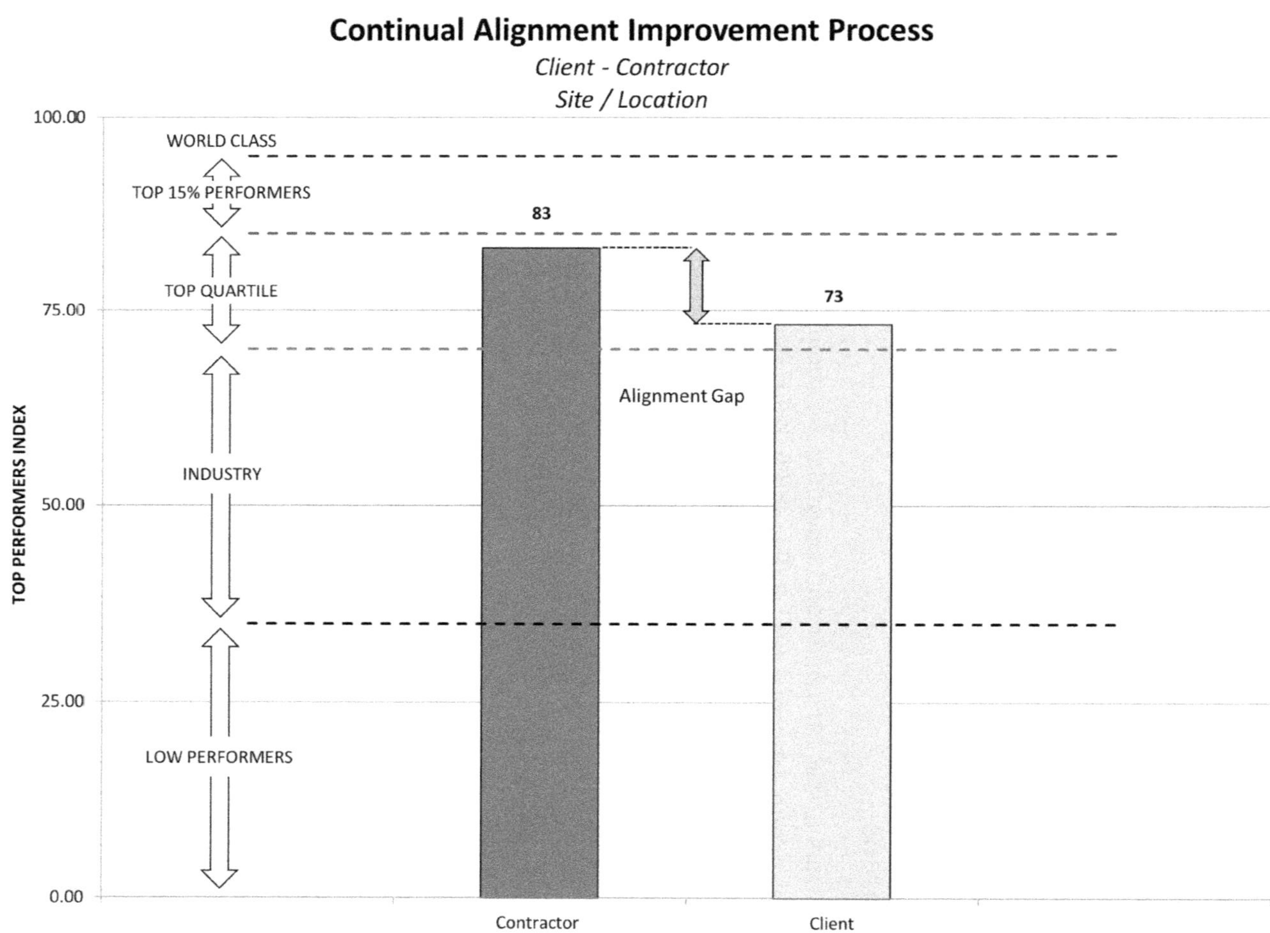

FIGURE 13.1 CAIP overall scores

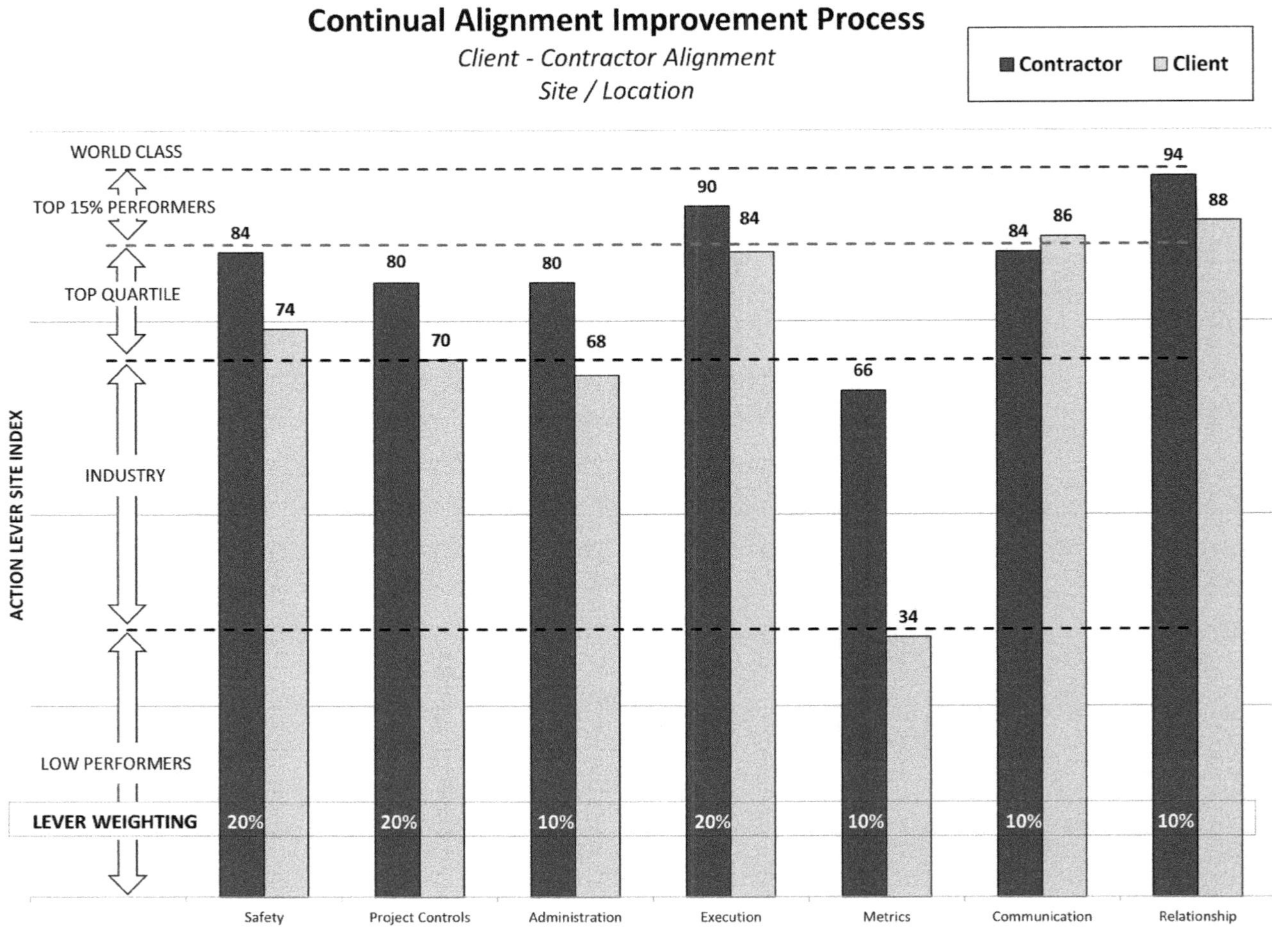

FIGURE 13.2 CAIP tier scores

- ❑ What does GOOD look like? What does GREAT look like? What does WORLD CLASS look like?

Define circuit alignment by the customer's and service provider's steering team

Step 3: Strategic objectives

Development of the strategic objectives to improve

- ❑ Agreed-upon strategic objectives derived from elements with building blocks and targets/goals
- ❑ Specific activities that need improvement
- ❑ Short-term versus long-term needs

Step 4: Cause-and-effect linkages

- ❑ An understanding of the positive and negative value for the customer and the service provider
- ❑ The impact on others (customer, service provider, EPCs/GCs, etc.)

Step 5: Metrics—development of the improvement scorecard

- ❑ Document the levels of achievement for each strategic objective.
- ❑ Assign accountability.
- ❑ Determine how to verify sustainability.

Step 6: Evaluations—results

The customer and the service provider work together to answer:

- ❑ Are our strategies working?
- ❑ Is the culture changing?
- ❑ Are we measuring the right things?

Step 7: Best practices

- ❑ Have we discovered a best practice?
- ❑ How do we make it work at all facilities?
- ❑ Is it the key to circuit alignment?
- ❑ How do we cascade it throughout the circuit?

Step 8: Continued improvement (period: every six months, once a year, etc.)

- ❏ Do we need/want to move to the next tier?
- ❏ Can we move to the next tier?
- ❏ Revise strategies.
- ❏ Repeat Step 2.

When conducted in a true partnership where mutually beneficial results are the goal, this CAIP results in a value-adding process that ensures that the following occur:

- Attain industry's best processes and people empowerment.
- Create proven performance and cost reduction strategies.
- Capture best practices from a diverse customer base.
- Implement a win-win approach to contract strategies and alliances.
- Gain alignment of management teams to achieve specific desired outcomes.
- Capture innovative ideas and technologies.
- Create commitment to excellence and continual improvement.

We have now ensured alignment with you and your selected partners to know the next steps to take together to prevent the return of any more maintenance insanity as long as we hold each other accountable to achieving our goals. In the next chapter, we will investigate several assessment tools that will tell us where we are so we can decide where we want to go.

CHAPTER 14

Assessment Processes and Cold Eye Reviews

You cannot change what you do not acknowledge. Stop making excuses and start producing results by evaluating your maintenance insanity and selecting the things you want to improve. The purpose of any assessment is to determine "where you are" relative to "where you want to be." If you do not know what "good" looks like, you may need to bring in outside help or subject-matter experts to help you develop your vision. The benchmarked tiers showing levels for selected categories at www.maintenanceinsanity.com allow you to get a feel for your present condition. The objective of an assessment is to help your site, area, or work group meet business expectations and drive ongoing improvements by identifying gaps in performance. Assessments are also great tools for communicating your new vision and for identifying best practices throughout your organization. By setting priorities, the findings from an assessment can also guide investment and leverage resources for targeted improvements in work processes, systems, and tools. A key value of the assessment will be to ensure everyone understands the vision so that any current practices that are not in alignment with the vision can be curtailed and so that any incremental activities or initiatives do not propagate movement in an undesirable direction.

You can help with the assessment to ensure objectives are met by previewing the assessment content and providing feedback on anything you feel is missing to be able to evaluate your site. Encourage appropriate preparation of any requested documentation. Make sure people understand that the highest-assessment-level score indicates vision or the desired state and may not be attainable or desired for your site. The lowest-level score indicates minimum requirements and shows the starting point. Your position on the scoring scale identifies your opportunities. Highlight to all participants that an honest assessment (gap analysis) will provide the most value. After your initial review, provide feedback on possible enhancements to the assessment. An assessment is to bring into focus an optimal path forward for improving your plant reliability and maintenance productivity for improving your bottom line. The assessment provides both quantitative and

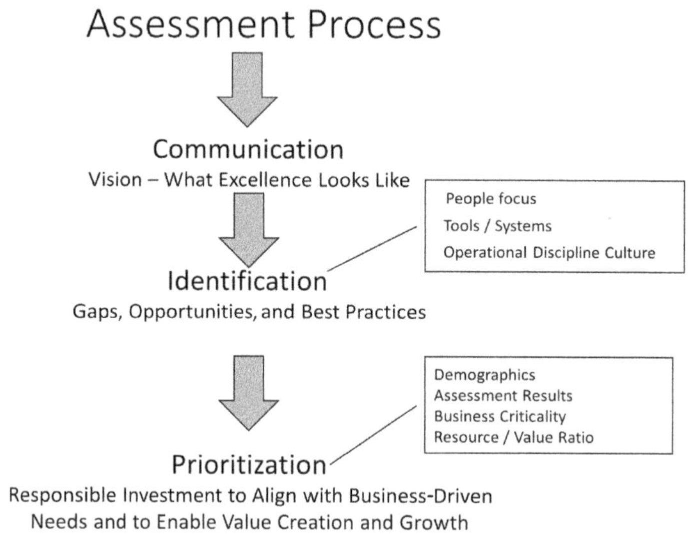

FIGURE 14.1 Assessment process flow

qualitative feedback on the strengths and weaknesses of your current maintenance and reliability program. Through metrics or key performance indicators, you can identify gaps in your current processes and determine the benefit of closing those gaps to deliver a more efficient, productive, and profitable strategy for operation. The simple assessment process flow shown in Figure 14.1 provides an overview for the assessment process no matter what tool you use.

After the results are compiled, small variations will not be reviewed unless questions exist between the team members for understanding. Significant variations will be reviewed and will require group discussion to determine the consensus score. Low scores represent gaps that can be improved. Top scores indicate best practices and should be documented for sharing with others. A draft of the final report should be sent to the site or area manager to review to ensure that the wording of the gaps and best practices is appropriate. The final report will include all files and data with a summary and recommendations. Best practices should be stored on a common access server in a library to share throughout the organization. You will need to prioritize the gaps for your area and work on the most important gaps with available resources. You are in control of how the assessment results are used but can ask for support as needed from outside resources. Set a high priority if the identified gap is foundational to ensuring that the current value from your department is not at risk. Of those nonfoundational gaps (those where enhancement could enable growth or create additional value), the team can rank them in order for addressing. Consider the following questions for setting your ranking:

- Do the business objectives/needs for my department align with addressing the gap?
- What is the value for improvement?
- What level of resources is required to address (consider available best practices that could be easily deployable)?
- Does the resulting value to resource ratio suggest this is a high priority?
- Will I have the resources to address within the next year?

Figure 14.2 gives an explanation for one type of scoring.

Here is an example of an executive summary for a maintenance assessment for a medium-size chemical company. See if any of these comments ring a bell for you.

Score	Level of Implementation	Attributes and Evidence
5	Optimizing (Best Practice)	Effective and efficient. Visible continual improvement culture / efforts in place. Systematic application of lessons learned and best practices to improve performance.
4	Managed	Documented and effectively implemented. Consistently following programs, procedures, and systems.
3	Implementing	Documented. Implementation ongoing but not always effective. Established programs, systems, and procedures and are actively working on implementation effectiveness.
2	Program Development	Some evidence that the practice exists with limited documentation. Aware of opportunities and are developing programs and procedures.
1	Awareness Building	Practice is inconsistent or ad hoc. Reliance on key personnel. Reliance on initiative.

FIGURE 14.2 Scoring explanation

Executive Summary

Company A's Dallas plant has been feeling the same pressures as the rest of the chemical industry for the past several years with increased costs, reduced margins, and lower profits. The management of the company is dedicated to getting costs out of the system and increasing profitability wherever possible. There have been several employee reductions in the past five years that have left the remaining employees unsure of the future of their jobs. The employees also know of discussions with other companies interested in outsourcing the maintenance program. In the areas of the plant we visited, there is sound process knowledge and there remains a willingness to work.

The relationship between maintenance and production seems to be good. Mechanics are assigned on a semiformal basis to individual production units, which give them a feeling of ownership. This relationship is somewhat less firm than it was prior to the last round of layoffs. The mechanics are likely to be pulled elsewhere. Production operators always like to have a mechanic handy, in case something goes wrong. This is not, however, the most cost-efficient way to distribute the mechanical and instrument labor forces.

The maintenance organization at Company A is divided into several small groups that require a number of people, primarily the maintenance coordinators, to function in a variety of roles. Management has a vision but no plan for a more reliable operation with reduced failures, lower cost, and increased production, and it is unclear what initiatives should be undertaken in order to achieve these goals. The goals for the production organizations are pretty clear and straightforward and have established measures that are reviewed regularly. There is no corresponding set of goals, plans, or measures for maintenance. Maintenance has no vision, strategy, or objectives. When asked, most maintenance employees were unsure of what was expected of them and what the company was trying to accomplish.

In essence, the plant has built a very good basis for a modern, effective, and efficient maintenance organization. Implementation of recommendations found during the assessment will allow for the creation of a maintenance organization meeting the plant business goals with minimum cost. The experienced maintenance workforce is one of the biggest assets of the organization. However, one drawback is that the workforce is aging. This has been accelerated by the reductions from the ranks of the younger employees over the last five years. There is not a very thorough training program for new operators and mechanics. It is assumed that employees are hired with the basic skills and are then put with an experienced employee for a short time for on-the-job training. This is not optimum since the new hire will often miss learning how to handle operations or problems because equipment was not running at the time. In addition, there is always an argument about exactly how equipment should be set up and run from crew to crew. These issues need to be resolved and the best training method determined.

Based on the assessment findings, the following are the issues that must be addressed first, for three reasons: (1) Each can be completed in less than six months to the point where savings will be evident. (2) Of the elements that can be done quickly, these have the highest payoff. (3) These are all foundational building blocks for the long term.

Maintenance Strategies

Maintenance strategies must be developed and initiated. These strategies will address how you will do maintenance. They will be the guiding expectations for all maintenance employees and must be understood by production employees.

Management Leadership and Support

Management must clearly and openly support the changes that will take place. The plant manager is the keystone that holds the program together. It is the plant manager who must visibly support and drive the expectations and strategies set forth above. He or she must lead the Site Management Team to link with the Site Maintenance Leadership Team to know how all are doing with their implementation plan.

Contract Management and Utilization

Someone has to get a good grasp on contractor use. Limits should be set based on the amount of critical work to be done and the maximum number of contractors to be used. An individual must have responsibility for this control and should be the only one who calls contractors. Guidelines and procedures can be established for 24/7 coverage when the contract coordinator is not available. The coordinator will review and question each call-in of contractor resources.

Workforce Management

There is little flexibility for sharing maintenance resources with the current organization. We propose to address this gap with a new organization.

Planning and Scheduling

One of the quickest ways to get gains in a maintenance organization is by moving to planned work. Wrench time by the mechanics can be doubled by removing most of their dead time looking and waiting for parts, tools, and safety equipment and delays from equipment shutdowns and incomplete locking and tagging. This is done by focusing a few individuals on some basic planning functions and not releasing the work for scheduling until the planning is done. The proposed organization also is essential in achieving this result.

Maintenance Improvement and Reliability

If anything is clear concerning maintenance work, it is that the only way to effectively reduce maintenance work is to take work out of the system. Another way of saying this is that eliminating the cause for the work is the best way to prevent the work. Repeated seal failures? Get the right seal for the job. Constant unexpected line leaks? Do some predictive measurements to predict them before they happen. To do these things, you need to have a detailed maintenance improvement and reliability program in place that defines what preventive and predictive maintenance must be implemented on a data-driven basis. Root cause failure analysis must be done on failures to get down to the latent cause, not just what part broke. To do this, you need to reinstate the two reliability engineer positions.

Maintenance Organization

This is the foundation for all other recommendations that must be done. While no one can ever reorganize for success, your organization can surely set you up for failure. We believe your current organization falls into this last category. While we were fairly comfortable proposing a new organization, it may seem fairly presumptuous assigning names to the positions. We based suggested names on our interviews with a number of your people and offer the interviews as a starting point for discussions. We certainly expect your input and approval prior to any assignments.

All Assessment Findings Should Be Justified by Potential Financial Impact

We have attempted to estimate the financial impact of the maintenance improvement program on the overall plant business. We did not have the data required to calculate the impact of spare parts reduction or contractor reduction, and so we built these into the overall maintenance savings.

In summary, we believe that there is potential for savings of approximately $1.7 million annually in maintenance expenditure in the first year and a total (cumulative) savings of $5.1 million after three to four years if the program is fully implemented. The potential monetary gains can only be achieved if all the recommendations are implemented to the full extent. There are additional gains to be made from plant uptime improvement. These normally exceed the gains from maintenance cost, but we were not provided with the information to calculate these.

In addition, safety was not mentioned as one of the initial target areas merely because the economic impact did not make the cut. However, there would be a great increase on the emphasis for safety and an equally great reduction for the tolerance of unsafe acts. The current record is clearly not acceptable and must be improved. Planned jobs are safer because you know what to expect and hazards have been identified.

The savings potential is an estimate based on the data provided by the plant and the assumption that the recommendations contained in this report are implemented by the maintenance improvement program. Additional savings on spare parts management and production uptime are likely but could not be calculated due to lack of data.

Assessment Results

The assessment process is made up of data analysis, plant interviews, and on-site observations. The assessors scored 21 elements by determining the current state of performance in 7 to 10 subelements for each area, arriving at an overall score for each element from 1 to 5.

In general, a 1 indicates a strategy or process not being practiced and a 5 indicates world-class performance. In addition to the current state, the assessors indicate the score that Company A could reasonably expect to achieve in one year and long term. No more than five to seven elements are picked for improvement in the first year, these being the ones that would give the greatest return in that time period.

The spider chart shown in Figure 14.3 represents a view of current and future states. In this case, only those elements targeted for initial work show their one-year scores. You will also notice that very few of the elements show a future state at the 5, or world-class, level. This is because it does not appear to be practical or prudent for Company A to spend the time and resources to try to achieve closing the gap to that point. As can be seen, the biggest gaps are in the areas of planning and scheduling, maintenance organization, equipment conditioning monitoring, maintenance improvement/reliability process, and maintenance strategies.

Maintenance Assessment Overview

Purpose of the Assessment

The purpose of this assessment process was to review the existing organization and provide comments on existing maintenance systems, practices, and procedures. It was not the intent to review the performance of any individual employee or group of employees. The intent was to:

- Provide short-term recommendations focused on the development and implementation of a maintenance improvement plan (maintenance strategy).
- Provide longer-term recommendations to assist continual improvement in asset management and workforce development.
- Provide estimated potential savings focused on the developed and implemented maintenance improvement programs over one-year and three-year periods.

The specific objectives were to:

- Carry out a comprehensive maintenance assessment of the organization and provide comments on existing systems, practices, and procedures.
- Provide long-term recommendations that would assist in the development and implementation of continual improvement in asset management.
- Identify opportunities for continual improvements toward best practices.
- Review the operating structure with respect to best practice organizations focused on core business requirements.
- Review workflows and procedures with respect to the value chain of the business and to identify non-value-added work.

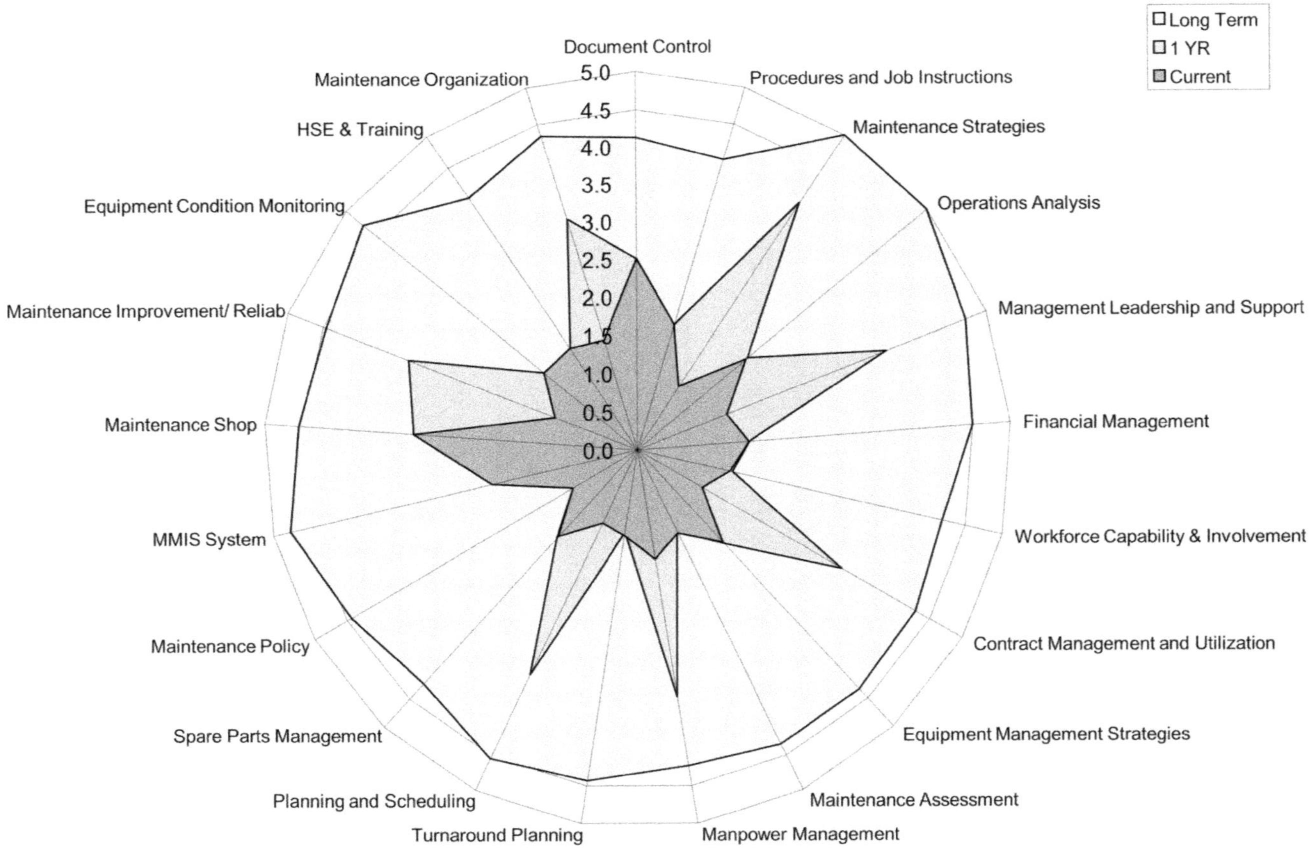

FIGURE 14.3 Current, one-year, and long-term radar chart

- Review and comment on the planning, scheduling, and execution of work targeted for planned maintenance windows.
- Review, comment, and make recommendations on the planning and execution of major shutdowns.

Maintenance Assessment Process Description

The maintenance assessment system used was developed over several years of using a number of different types of processes, from those requiring a tremendous amount of data to be gathered and a lot of time on the part of those being assessed to a self-assessment process that left a great deal of uncertainty about whether the assessed understood the questions and the assessors understood the answers. The process now being used removes that uncertainty. By conducting interviews with a cross section of individuals, from every area that maintenance impacts and from all those areas that impact maintenance, it is possible to get a very clear understanding of the processes and strategies being used by the client. One can see the strategies not only as they are intended, but as they are actually being performed. It is thus possible to see disconnects between intent and practice that result in improvement opportunities. In addition, the assessment reveals areas where no strategy or process exists, areas that may be foundational to a good reliability program. It does this while at the same time minimizing the time required from the client's employees.

It is our belief that a comprehensive reliability program is the only one that will truly succeed over the long term. There are many components that make up a successful program. Very few can reach their maximum benefit without being combined, at least in part, with others. It is the blending of these elements that we believe has led to our success in supporting numerous plants. Many plants have some or even all of the components but use them in an ineffective manner. It is the goal of the assessment to guide the development of the right strategies, processes, and practices for each site, taking into account the plant's own strengths and needs.

It was for these reasons that the 21 elements were selected. They cover all the components that go into developing an effective comprehensive reliability program. The elements are

1. Maintenance organization
2. Health, safety, and environment
3. Maintenance support
4. Equipment condition monitoring
5. Continual improvement/reliability process

6. Maintenance shop
7. MMIS system
8. Maintenance process
9. Training
10. Spare parts management
11. Planning and scheduling
12. Turnaround planning
13. Workforce management
14. Maintenance evaluation
15. Equipment management strategies
16. Contract management
17. Workforce capability
18. Financial management
19. Management leadership and support
20. Production analysis
21. Document control

While each of the elements is important, they do not all carry the same weight, either relative to each other or from plant to plant. Size, organization, unionization or not, product produced, all can make a difference in the value of a particular element. The assessor must determine, through the interview process, the relative merits and potential for gain of each element with respect to the client. The complete assessment tool is available at www.maintenanceinsanity.com for your facility to use when you are ready to begin improving.

The assessment procedure includes these steps:

- A preassessment questionnaire is sent to the primary contact ahead of the visit to obtain background information to familiarize the Assessment Team with the business. Bear in mind that you may not receive a complete set of data.
- Based on the information, the Assessment Team develops an interview timetable, identifies key participants, and forwards its work for approval. Generally, the first two days will be firmly scheduled with the remaining time only tentatively scheduled.
- The maintenance assessment starts with an introductory meeting conducted by the Assessment Team at the plant's premises. All key people should be present and as many of those assigned to be interviewed as possible.

- After this introduction, the Assessment Team members walk through the plant and maintenance facilities to familiarize themselves with the general operation and geography of the business.
- The Assessment Team conducts the interviews with the participants. This will normally take three to four days.
- On completion of the interviews, the Assessment Team will conduct an "exit" meeting with the key participants for a general discussion of any specific outcomes from the review process and will outline any major observations and immediate short-term recommendations.
- The Assessment Team leaves to evaluate the interviews and observations made, finalizes comments and recommendations, and generates the maintenance assessment report.
- The report is sent to the client for review approximately one week prior to the Assessment Team's returning to present the results and answer any questions.
- Following the presentation, the assessors will meet with the key contacts to discuss the path forward.

Scoring Charts, Comments, and Recommendations

After conducting the interviews, the assessors score the individual elements by determining the current state of the client in from 7 to 10 subelements in each element, arriving at an overall score for the element of 1 to 5.

In general, a 1 would indicate a strategy or process not being practiced, and a 5 would indicate world-class performance. In addition to the current state, the assessors indicate the score the client could reasonably expect to achieve in one year and long term. No more than five to seven elements are picked for improvement in the first year, these being those that would give the greatest return in that time period.

Probably the most valuable part of the report is the comments and recommendations section. It is here that each element is looked at in detail. There is a summation of the findings: things that are going well and things that are improvement opportunities. For those with opportunities, there are recommendations regarding the steps that need to be taken to realize the improvement potential.

This entire assessment process is shared at www.maintenanceinsanity.com. The tool (complete spreadsheet including scoring and feedback graphs) containing the described assessment process can be customized to fit your needs.

Cold Eye Reviews: Having Someone Not from Your Site Evaluate Your Processes to Get a Fresh Perspective

When you are working with a cluster or circuit of sites under the same ownership, a good way to benchmark your performance is to partner with your maintenance or contractor service provider to do joint deep-dive visits to capture a cold eye perspective among the different site cultures. There are usually plenty of surprises when the corporate standardization is actually observed in practice.

The objectives for a deep-dive visit include identifying barriers to efficient work execution, evaluating time and material or unit rate work processes against established framework protocol, identifying opportunities to increase T&M or unit rate utilization and improve productivity, and taking learnings back to other sites for improved productivity. A cold eye review sees things that the host sites look past each day and allows them to make quick wins from these learnings.

To ensure an efficient use of the team's time and to gain maximum benefits, the team needs to be selected from the client and maintenance contractor site leaders (pairs from each site) to coordinate schedules for availability. The following list of required documents will give the team members a head start on what to expect and the best ideas on where to focus their attention to truly understand the visited site's planning and scheduling work processes.

Deep-dive documentation that is to be reviewed by the team prior to arrival at the visit site should include

- The current and prior week's work schedule
- The distribution list for the weekly schedule
- Lead time for work requests
- An example of a good job package given to the crews
- Scoping sheets
- Examples of job estimates by planners
- Barrier or delay data
- Roles and responsibilities of key P&S positions
- Utilization trends
- Procedures and work processes
- Feedback given to planners
- Productivity trends
- Organizational charts for client and contractor

- A snapshot of the head-count data by area
- Break-in criteria and consequences

Create a visit agenda with an interview schedule for coordinating personnel and spaces for meetings and site visits. Separate your team members into three focus areas: planning, scheduling, and execution. Here is a sample schedule from a typical Day 1:

Day 1

- **7:30 a.m.** Arrival "Meet & Greet"
- **8 to 9 a.m.** Kickoff Session
- **9 to 10 a.m.** Team 1 Planning—Client Routine Maintenance Planner, Team 2 Scheduling—Data Review, and Team 3—Client T/A Execution
- **10 to 11 a.m.** Team 1—Contractor Routine Planner, Team 2—Contractor Routine Scheduler, and Team 3—Contractor Majors Foreman
- **11 to noon** Team 1—Data Review, Team 2—Client Routine Scheduler, and Team 3—Contractor Routine Foreman
- **Noon to 1 p.m.** Lunch
- **1 to 2 p.m.** Team 1—Client Routine Leadership Interviews, Team 2—Scheduling Group Discussion, and Team 3—Client Execution Foreman
- **2 to 3 p.m.** Team 1—Contractor Routine Foreman, Team 2—Field Observation Visits, and Team 3—Field Verification Visits
- **3 to 4 p.m.** Team 1—Contractor T/A and Majors Planner, Team 2—Client Majors Planner, and Team 3—Continue Field Verification Visits
- **4 to 5 p.m.** Team 1—Client Majors Planner, Team 2—Contractor Majors Foreman, and Team 3—Contractor Routine Foreman

The complete deep-dive "cold eyes" assessment tool and process including the interview schedule is shared at www.maintenanceinsanity.com for your utilization.

Planning Team typical questions include

- What is the process for executing a job? Who initiates the request? Who approves the request? How is it documented?
- Does the site have a defined, expected method of work notification to the contractor? What is it? Is it consistent across different work groups?

- Do the planners like the work notification process? What improvements would they like to see?
- Does the input cover all work groups and computer systems?
- Pick three work orders (various crafts). Was the notification made consistently with the site's defined method? How far in advance of the last acceptable completion date was the notification made?
- Pick three work orders from the current week's schedule and review the scope provided with the work order. Does the scope include the appropriate detail?

Scheduling Team typical questions include

- How are pushed jobs rescheduled? How are the holes created by pushed jobs filled during the week?
- Obtain a copy of a sequenced schedule for routine, majors, and program. Is the sequence laid out by zones? Does the sequenced schedule represent a 40-hour week for all crew members scheduled? Does the sequenced schedule include "filler" jobs?
- Review the schedule compliance. Are jobs on the schedule worked? How many jobs are worked that are not on the schedule?

Execution Team typical questions include

- Pick a break-in from last week. How was it communicated and documented? Who determines which crew will respond to a break-in? How is it decided which jobs will push out to accommodate the break-in?
- Is approval given for temporary material storage in the areas to be ready for the start of work?
- Review the barrier data for delays. How is the data collected? How is the data used?

In addition to the learnings based on the team's observations and interview answers, some general gains from this type of site visit and discussion include these findings:

- An understanding is gained of the planning and scheduling work processes across different sites.

- Report-out documentation makes great benchmarking information for sharing.
- Data reviews identify needed changes at other sites as well as the visited site.
- The audit team members should participate in regularly scheduled meetings to see the true interactions between the client and contractor.
- The material coordination people need to be included to get a better overview of all activities.
- All the site leadership needs to be involved in the kickoff and wrap-up sessions, conduct short interviews during the site field visits to gain a different perspective, start early to see the crews' "first starts," and visit contractor facilities to see their setup and space limitations.

The discussions by the people from various sites for both sides provide the greatest value when they see and understand what is done better, worse, or the same compared with their own sites. Everyone comes away with a new perspective and many improvement ideas.

The best way to show status and growth from a follow-up visit is to have a scorecard that is presented and explained during the exit wrap-up meeting. It also provides consistency when comparing feedback from all the visits. The scorecard for the deep-dive assessment provided at www.maintenanceinsanity.com gives you the general idea of what to capture. The components of the scorecard are evaluated for the ranges of 0 to 2 unacceptable, 3 to 5 minimum level, 6 to 8 average level, and 9 to 10 best practice. Word descriptions are provided for each level on the score card example at www.maintenanceinsanity.com.

The components are

Planning

- Work requests
- Zone assignments and input
- P&S input and approvals

Scheduling

- Schedule in use
- Framework steps followed

- Contractor's sequenced schedule
- Material handling

Execution results

- Percent schedule compliance
- Percent total labor-hours worked on schedule break-ins
- Productivity (estimated hours versus actual hours)
- Contractor RSCH (ready to schedule) backlog

Total score

Now that you know how to assess your state of maintenance insanity, you only have to look through the list of opportunities and pick what you need to work on first. You must "name it to claim it." Not knowing precisely what you want is not okay. Be very specific and focus on the goals you set. If you cannot measure it, you cannot improve it. Create a vision to grow toward.

The next chapter gives you a planning process for forecasting future needs through sales and operations planning to provide you with a structure that will prevent sliding back into the maintenance insanity trap.

CHAPTER 15

Sales and Operations Planning

When you are stuck in maintenance insanity doing the same things over and over, not only do you get the same results, but you do not know how to forecast and predict something different. Sales and operations planning (S&OP) is a process designed to promote increased visibility and optimization of forecasted demand for assets and resources with planned supply, ultimately giving your company a competitive advantage in managing your business. By combining company vision and strategy into one unified operating plan, the S&OP process also serves as a critical communication tool throughout the organization. It takes individual and group support at all levels to make S&OP an effective component of your management process. S&OP can benefit every aspect of your business no matter the type or complexity. It works for service organizations and entire corporations.

S&OP is a methodology for managing and operating a company's business effectively. It is fundamental to growing the business. This is truly a journey with ongoing progress as you work through the established steps.

Throughout the chapter, we will use the feedback from a benchmarking visit with a chemical manufacturing company that has been very successful with its S&OP process. S&OP defines the process the company follows and the results achieved. The same person has been the lead for S&OP from the initial development. The person reports back through a Steering Team composed of top management. The initial focus and key to success is to accurately forecast the demand. This step allows the creation of Planning Teams to obtain the needed supply. They are focused on customer product demand, raw material, and the company's manufacturing capability (supply). S&OP is a foundational process to do the company's business.

The company's secret is that SAP links all its data together, the employees are all using a common process and tool, everyone has the same vision, and S&OP provides a consensus plan for daily management. The business units' own S&OP and supply chain facilitate the process for the entire company. The company transitions new acquisitions *immediately* for S&OP and to their SAP data system. Much of the transition actions take place even before the final agreements are signed, so the process is in place from Day 1 of new ownership.

The demand forecast was hard to drive at first, but the accuracy is the focus metric for tracking improvement and for gaining rewards and recognition (monthly and quarterly). Demand forecast is for products with accurate inventory critical to drive better yields (productivity), schedule shutdowns to fit demand, and allow you to react to unplanned outages and occurrences. Anything you do to better understand demand improves all other measures because of the multiple levels of services that are affected. If you can increase accuracy even 1%, you have moved cash from holding inventory to growing your business. One of the chemical company's main locations had a complete plant outage due to a total electrical failure that was safely handled without loss of customers or significant impact. The test for supply loading is if it meets your demand without interruption due to lack of inventory. For S&OP to work, you must understand your supply and demand.

The last purchase made by the company was a turbine oil business, which will enable the company to become a supplier to the airlines. As its first priority, the company is fully committed to transitioning this business to S&OP and SAP.

The chemical company started its S&OP journey in 2007 working with an industry research consulting company. The initial focus was to get a true picture, with product demand at the top of the pyramid and with assess, diagnose, and correct as the steps below. It wanted to better manage its inventory to show cash impact by making the inventory available for other uses as needed. The company ships to over 100 countries and needed a buffer for the products it makes. The length of each supply line sets the inventory levels and where to keep the inventory. The desire is to minimize inventory based on demand accuracy. The board of directors sees this as the way to grow the company at minimum cost. By its choices and commitment, the company owns S&OP (not the consultant).

Its road map was to be the innovator for the supply chain with a five-year plan to enable improvement, add technology, leverage capability in its people, be driven by leadership commitment, and achieve a culture change throughout all its companies and locations. It now has 7 supply chains by business and 41 S&OP Teams. Each of its businesses has unique needs. Some focus on agility for product mix; some focus on responsiveness with one having a 15-minute window for delivery and still maintaining efficiency for being the low-cost provider. The company picks the most important focus for each product line to craft the supply chain strategy. Short- and long-term goals are all linked through S&OP to improve profitability.

This company's S&OP integrates and synchronizes efforts through a six-step process:

1. Discover new capacity and new demand.
   - By using a funnel concept, the company is able to filter through and focus on its most profitable demand scenario.
2. Manage demand.
   - Through consensus meetings, the company is able to come up with a demand forecast that all business units believe in because they had input into it.
3. Do supply planning.
4. Do a financial evaluation and reconciliation.
5. Create a consensus plan.
   - All business units are engaged and actively participate in creating a plan based on the agreed-upon demand. Decision makers must be present at the meeting to finalize the plan.
6. Publish and communicate the plan.
   - None of the above steps will be worth the effort until the plan is documented and communicated so all areas of the business are headed in the same direction.

This requires linkage of demand, supply, business initiatives, resources, and financial management. The product planners use it most. There is now only one demand plan where there had been six. The consensus plan is the only one now. This is used for everything from inventory control to supply chain management to planning future financial budgets. Demand causes you to be creative on supply.

The company's initial time focus was for 1 month out. Next, it moved to a 3-month, then 6-month, and now 24-month forecast. The further out you forecast, the more options you have to meet demand. With the financial linkage now, the projections have less than a 1% adjustment. The company sets targets and measures performance against the (only one) plan. Accountability for forecast accuracy is for 3, 6, 9, and 12 months but not for the second year (the most important value gained is the best guess where the trend is). The company does not hold forecast accuracy and accountability to the same level for the extended periods of time (3- and 6-month forecast accuracy is held to a higher standard than 9- and 12-month forecasts). This keeps management from overreacting. This is a maturity issue.

The company now recognizes and rewards the most improved forecasting group or person and the ones with best accuracy each quarter. The company also now has full-time forecast managers focused on current and future growth (but this took a long time to sell). The managers are charged to make 1% to 2% improvement on an ongoing basis. These people handle the drama (each day). For the company consensus meetings, everyone participates from upper management through all functional groups and operating units. The business unit leaders show their demand and plans to address it, and others share options to determine the best trade-offs. Financial uses a company-built bolt-on tool for SAP so that all the information is connected.

SAP made a huge difference by finding inventory "we did not know that we had" due to all the different computer systems that had been tracking it. On a scale of 1 to 5, the company was assessed at 1.2 when it started. It has taken it seven years to get to level 3.

The levels are

1. React
2. Anticipate
3. Integrate
4. Collaborate
5. Orchestrate

The movement along the scale may never allow you to get to level 5 and stay (and you may not want to). The focus of Step 2 is to get all the groups together and work on trusting each other as well as building teams. It took the company four years to get to this point before moving on to Step 3. In Step 3, you are able to trust the consensus demand and use it in all areas of your business.

You must celebrate your small wins to keep moving along the curve and making progress. Achieving your first consensus forecast (by all parties) is a major milestone. Work with the end in mind to set steps and celebrate reaching each one. At Level 3, the people in the company now trust each other, and all use the same forecast. They do "what-ifs" to stay on the same page and give financial results for each scenario. This gives them full visibility into all their options so they can choose what steps to take to effect business growth. The company has found that 67% of companies doing S&OP never get past Level 2.

The value creation the company expected from S&OP was a two-to-one payback. Six to seven times the value has been added based on new business that it has led to. A quote from one of the executives is "Growth is not easy and growth is not cheap. It will cost to make efficiency gains and requires dedicated resources." Do not start the S&OP journey if you are not committed to see it through. It takes seven years to fully change your culture. The chemical company is just now arriving—it took the company four years to see the major successes. The S&OP Steering Team is made up of the highest levels from the different functional groups that are united in their feedback to the board.

S&OP wins because it is a better process and builds the foundation for future success. It is an integrated consensus-business planning tool. You can make the investment in growth or continue to manage drama. The chemical company's goal is "Don't be stupid on purpose." You must walk before you run by getting the basics in place first.

Start with your business strategy and link it to your S&OP process; then allow feedback to modify the strategy as you mature. Operational execution links supplies, resources, plan/make, delivery, and customer satisfaction. The common target is for consensus demand to identify what business to go get. We must all have the confidence to pull the trigger and move forward.

The company is on a monthly S&OP cycle with a documented communication plan. It is critical to hire and retain good people with established roles and responsibilities defined down to their decision rights. The management recommends that a service company client pick a pilot area to drive toward a consensus demand forecast as the first achievement. The client started with one North American product and created a scorecard for 3-, 6-, 9-, and 12-month forecast accuracy. This was shared monthly throughout the client. The year-over-year progress shows the pulse of the client. It is critical to get all the demand in one system. The chemical company also uses Tableau as one of its tools for sharing its data. A consultant may be beneficial to share best in class and a framework for S&OP.

It was important to this chemical company that the client be a partner since the client was viewed as part of the company's own success. S&OP works for the construction industry, oil and gas, and chemicals for skilled hard and soft crafts for nested accounts and large projects. The company has also seen a positive impact on its terminal operations by its strategy-driven approach with input from the contractor's assessment.

The company wants to partner with all its service providers on a linked S&OP journey. It believes that S&OP helps to grow its business to make more money. Each participating company must ensure ownership of S&OP at all levels of the company to be successful.

As its starting point, the service provider had multiple work processes for each business for the same services. Standardization is a critical first step to getting the basics in place. Getting and maintaining an accurate inventory is a must. The creation of the Regional Asset Manager Team and the Regional Talent Acquisition Owners Network Team was the first step to get a handle on the supply issues to meet the demand forecast from operations. You must incorporate the financial and business development groups into the overall consensus plan. Buy-in from the CEO and the S&O Team must break down silos and build commitment. You need upper-level management executives (CEO through VPs) as a main part of the steering effort to drive the needed change. You must use S&OP to make a difference and grow the company. The Network Team structure will handle the drama better and reduce the number of people now involved. Quantify (a computer inventory tracking software) for scaffold tracking, and the Tableau workbook structure will be the single source for demand and supply information for the consensus plan. You must pick a focus area (select a single region) as a pilot to drive all these processes and create your first consensus plan for execution.

When you are starting from scratch, you must have some method of forecasting future needs (business development and estimating with field relationships with owners) to be competitive and to stay ahead of demand and growth. Use S&OP as a communication tool between shop affiliates and company sites allowing the S&OP Team members to serve as facilitators for the interactions with all the functional groups. These include:

- **Accounting.** Forecasting budgets as well as cash flow
- **Supply chain.** Forecasting purchasing needs in advance to leverage price breaks
- **Staffing.** Forecasting workforce needs in advance to fulfill labor demands
- **Asset management.** Forecasting equipment and material needs in advance to leverage owned versus rental expenditures
- **Operations.** Sharing people and assets with enough lead time to be prepared for planned projects
- **Business development and estimating.** Forecasting future work with adequate lead time and accuracy to allow operations and affiliates to respond with planned solutions

Analyze the field data and computer system output for areas of strategic business planning. Make recommendations to operations to move assets and people from job to job utilizing your full potential. Make recommendations to the supply chain and staffing based on forecasted demand from the field sites and projects. A pilot is the best way to create and finalize desired functionality for the S&OP process and to show its value for management to decide the path forward use. You must fully populate the database with meaningful data from accounting and field input to drive corporate recommendations and strategy development.

One tool that can aid with information sharing and data collections is a web portal for forecasting to reduce lag time and manual compiling and loading errors. Create dashboards to give business insight, and track forecast accuracy to understand and improve your forecasting ability, highlight areas of potential business growth, highlight areas of weakness and improvement opportunities, and provide geographical views of results.

Together Figures 15.1 to 15.4 provide an overview of the S&OP process. Figure 15.1 presents the S&OP organization with roles and responsibilities, Figure 15.2 shares a

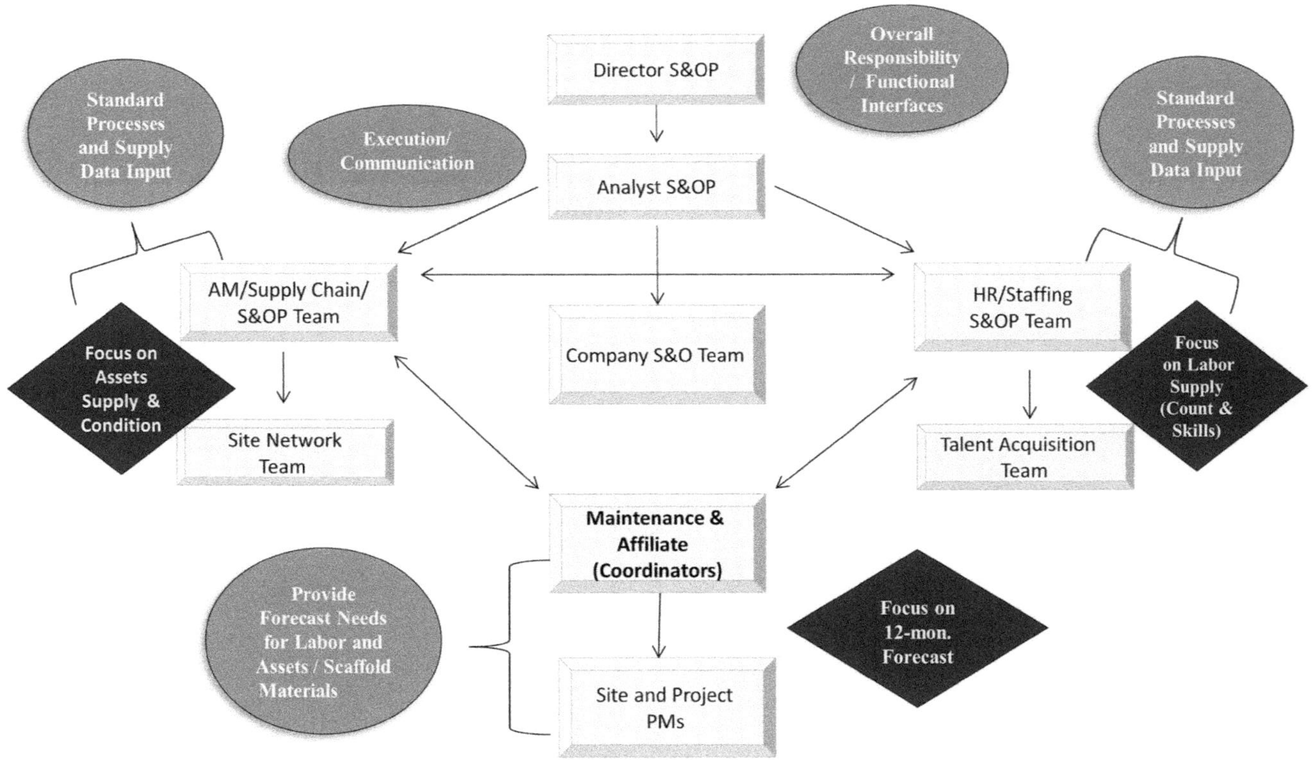

FIGURE 15.1 S&OP organization

FIGURE 15.2 S&OP inputs and outputs

simple diagram for the S&OP inputs and outputs, Figure 15.3 shows the S&OP process and timing for the monthly cycle, and Figure 15.4 describes the demand forecast process and data sources. These figures represent the key components for this process.

A properly implemented S&OP process will routinely conduct the following activities with associated benefits: It will review customer demand and supply resources. It will also "replan" quantitatively across a rolling horizon; the replanning focuses on changes from the agreed sales and operations plan. In addition, it will help you understand how the company achieved its current level of performance. Finally, it will focus on future actions and anticipated results and will monitor the execution of the company's strategies.

The most basic objective of an S&OP process is to align the supply of assets and resources with the projected demand. For the total company, projected demand includes the ongoing needs of current projects, the forecasted needs of potential projects, and internally driven needs for company-sponsored initiatives. The supply of assets includes owned equipment and materials and all rentals. The supply of resources includes general labor-hours, specific skills, supervisory capability, technical support, and costing/estimating expertise. S&OP provides a structured framework for making decisions regarding purchasing, hiring, resource assignments, staffing priorities, and pricing decisions.

The benefits to operations and shop activities are asset visibility and availability. It means you will have "good stuff" when you need it. It also means that there will be improvements in resource and asset quality and reductions in asset and labor shortages. In

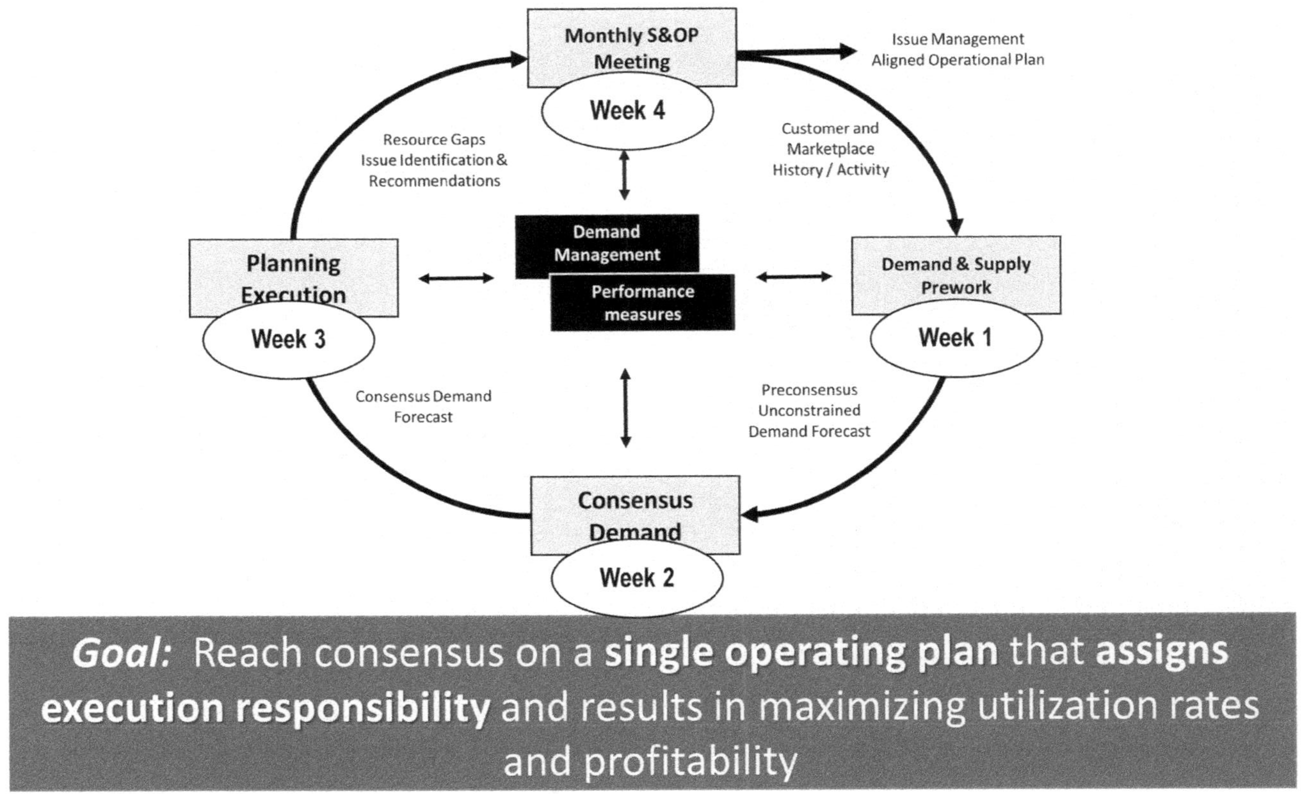

FIGURE 15.3 S&OP process and timing

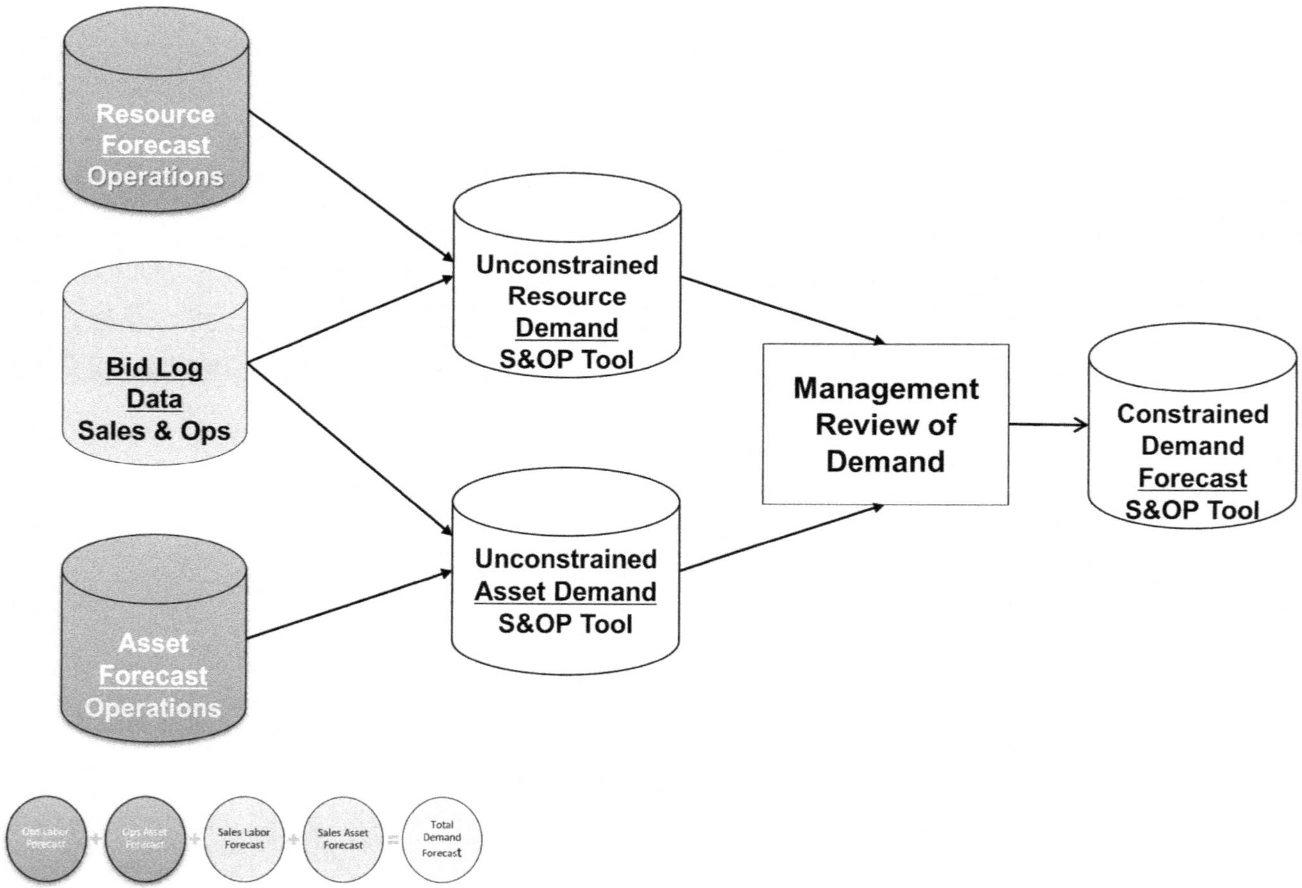

FIGURE 15.4 S&OP demand forecast process and data sources

addition, benefits include visibility to sales demand through awareness of future demand and the ability to proactively plan to meet that demand.

Specific deliverables for S&OP include

- A forecasting process that provides monthly, quarterly, and annual demand projections
- A supply planning process that evaluates the regional availability of work-hours, skill sets, and assets
- A standard monthly review process that identifies the gaps between the projected demand and available resources
- A gap-tracking and closure process to align the projected demand and resources
- S&OP planning workbooks and templates to support implementation and execution
- A trained S&OP manager to coordinate and sustain the monthly process

Anyone who has successfully implemented an S&OP process will tell you that it is not easy but is definitely worth the effort.

The next chapter addresses the handling of mergers and acquisitions to prevent someone else's maintenance insanity from infecting you.

CHAPTER 16

Mergers and Acquisitions

When you have a handle on your own maintenance insanity, you may get the opportunity to deal with someone else's when your company decides to expand. Growth may come to your company through the avenue of mergers or acquisitions. For maintenance as well as all other aspects of your company, the secret to a successful combining of companies or accounts is to have a smooth transition plan. Be prepared and do your homework prior to pulling the trigger.

An acquisition or merger deals with a lot more than combining companies. There will be many cultures and numerous individuals' concerns and worries. Be sure to incorporate the following change management principles when implementing your transition plan:

Principles of Leading Change

- Honor the past to understand the other side's maintenance insanity.
- Understand that leaders on both sides must demonstrate commitment to the change.
- Ensure there is continual communication. You cannot overcommunicate.
- Minimize ambiguity. Be specific on what is to change.
- Maintain and enhance the self-esteem of all individuals.
- Ask for help and encourage involvement.
- Provide support without removing responsibility.
- Commit the necessary resources.
- Listen and respond with empathy.

This process will be described for a major service provider contractor being replaced by another due to performance issues. You must have a proven site transition plan to provide a method to meet the initial staffing needs of the facilities and to enhance the work product of your own or your customer's operations while addressing the current incumbent employees and area community. Benchmark processes for improvements in efficiency and cost. This proposed approach is a disciplined, yet flexible, method that will assure an orderly and professional transition from the present owner or provider for either a single craft operation or a multicraft operation under the new banner at the designated facility.

Carrying out these steps will ultimately ensure your effectiveness in several areas such as:

- The seamless transition of work from the incumbent contractor to the new provider
- The development of an indirect site structure for multicraft services
- The service provider's relationship with the customer's plant personnel
- Implementation of the continual improvement process (presented earlier in Chapter 13)
- Implementation of the site business reporting process

Transition Team

Create a transition team consisting of a key service provider and site individuals to lead the transition process. The customer will identify individuals from its plant organization to participate on the team.

Transition manager	Facilitation of the transition process, action register management, transition timeline management, etc.
Administration	Contracts, organizational structure, policies, accounting, invoicing, financial services, etc.
EH&S manager	Environmental, health, safety, etc.
Human resources	Hiring, orientation, payroll, etc.
Technical support	Key performance management, engineering, work procedures, work processes, managed programs, etc.
Operations	Execution, day-to-day operations, facilities, work assignments, labor resources, equipment resources, etc.
Customer relations	Establishment of key linkages within the site

Transition Phases

This process recommends a transition plan comprising five phases, with each phase incorporating clear and detailed objectives to be achieved. The number of tasks included and the length of time involved in the transition are a direct result of the size of each site facility and the crafts to be transitioned, as well as other factors identified during discovery.

1. **Preplanning (2–4 weeks).** Upon the customer's formally awarding the contract to the service provider, you will begin the development of a Transition Oversight Team and the selection of the site-specific supervision and staff. This initial phase will include the development of a documented safety and management plan that will identify key milestones and objectives for an effective transition of staff, equipment, and field personnel. The customer's active participation in the development of this plan will be critical to ensure that site-specific concerns are identified and prioritized according to expectations.
2. **Site mobilization (1 week).** This phase involves the physical deployment of supervisory personnel and equipment to support the transition of work from the incumbent to the new service provider.
3. **Personnel transition (3 weeks).** This phase involves the hiring, skills assessment, knowledge validation, and employee orientation of field resources.
4. **Cultural assimilation (3–6 months).** This phase involves the implementation of site-specific safety work practices, field work processes, and reporting systems and the establishment of clear expectations relating to behaviors that will drive and support the service provider–customer partnership.
5. **Continual improvement (ongoing).** This phase represents the ongoing efforts by the service provider–Customer Management Team and the Site Owner Management Team to improve quality and productivity in complete alignment with the customer's business objectives.

Transition Highlights

As charged by the Transition Oversight Team, a detailed plan of the following site-specific objectives will be accomplished:

- Assignment of the Transition Team members
- Timeline for phase implementation and completion

- Alignment of the service provider's and the client's business objectives
- Alignment of environmental, health, and safety programs and processes
- Information system management interface
- Site computer systems
- Work assignments
- CSSM/backlog management
- Consolidation of contract structure
- Scaffolding, painting, and insulation (soft crafts)
- Mechanical (hard crafts)
- Organizational structure including indirect consolidation
- Evaluation and acquisition of on-site staff and resources
- Managed program development
- Site appraisal program
- Maintenance and service programs
- Transition, short-term and long-term key performance measures

Site-Specific Plan

The service provider will conduct a site needs analysis (discovery) in preparation for a successful site-specific transition plan with the customer's input. The needs analysis will provide critical information in the earliest stages of planning. The transition team will prioritize, assign ownership, and complete the task description as defined during the discovery meetings.

The emphasis of the transition plan development effort will be to create a plan and timeline with buy-in from both the client and the service provider. As part of the transition plan, the service provider and the customer will set a date for an official announcement to the plant and the community that the new service provider will be assuming responsibility for the contract services at the plant.

The announcement may create a period of high interest and concern. The new service provider will be prepared to handle inquiries from interested parties, especially employees of the incumbent contractors, in a manner that reflects positively on both the provider's company and yours.

The new service provider's general approach to the transition process for needed site personnel is to target the current employees of the incumbent contractor and incorporate them into the management processes for enhanced productivity performance and contract compliance.

Not all actions have to occur at the same time. Therefore, it is important to set a priority for action items based on when it is critical for them to be completed. The following table explains the priorities based on completion timing. The second table allows you to identify the owners' for these action items with back-up support names to be provided.

Plan Priority

1. Tasks that need to be addressed prior to the signing of the contract for services
2. Tasks that need to be addressed prior to moving onto the jobsite
3. Tasks that need to be addressed between the time you start moving in and the time you start executing work
4. Tasks that need to be in place once work starts
5. Tasks that need to be resolved during the first year of the contract

Process Control (Service Provider and Customer Representatives)	
Description	**Primary Responsibility**
AS	Administrative support (contracts, organizational structure, policies, accounting, invoicing, financial services, etc.)
HR	Human resources (hiring, orientation, payroll, etc.)
IT	Information technology (computers, phones, etc.)
EHS	Environmental, health, and safety
OP	Operations (execution, day-to-day site operation, etc.)
TS	Technical support (engineering, procedures, measures, programs, etc.)
BD	Business development

With the key players identified, you can proceed to develop the document shown in Figure 16.1, which is a sample transition action item plan.

Key transition items include the following steps:

1. Identify a site transition champion.
   - The service provider must have a site client leader in charge of the transition. This person can help with barriers and internal communications.
   - Help the champion will help with access to the key site individuals (IT, AP, HR, HSE, OPS, etc.).

ID	PC	PP	Task Description	Responsible	Target Date	Status
1	AS		Determine need and secure state licenses for jobsite.			Complete
2	AS		Review and document sales tax requirements for state.			Complete
3	BD		Set transition team kickoff meeting.			
4	AS		Secure LYB organizational chart and distribute to Transition Team.			
5	AS		Review and revise soft craft expectation.			
6	AS		Review and revise hard craft expectations.			
7	AS		Set formal contract signing date.			
8	BD		Setting ceremonial contract celebration.			
9	AS		Develop list of material suppliers with key names.			Complete
10	AS		Develop list of local vendors with key names.			Complete
11	AS		Set timeline for implementation of site appraisal.			
12	AS		Set timeline for implementation of site metric.			
13	TS		Determine and document LYB's requirements for subcontracting. Obtain LYB's approved vendor list.			
14	TS		Identify and document any potential "co-employment" risk considerations.			
15	TS		Identify and document problems with previous contractor.			
16	AS		Identify and document a list of LYB representatives. Distribute to transition team members.			
17	AS		Discover and document any LYB related politics to be considered during this transition.			
18	AS		Define transition team personnel and backups.			

FIGURE 16.1 Sample transition action items

2. Hold on-site discovery and planning meetings.
   - The service provider holds on-site discovery and planning meetings. This is envisioned as a full day of meetings with key site people. The goal of the meetings is to answer as many questions as possible from the transition plan details. The service provider will need access to a conference room and for someone to provide site client coordination for individuals.
3. Communicate major obstacles to the site client rep.
   - The service provider will communicate any major obstacles or gaps found during the discovery and planning meetings. Major obstacles must have a mitigation plan prior to the announcement.
4. Make contingency plans.
   - A contingency plan must be in place in case the incumbent demobilizes early.

5. Draft a communication plan.
   - The communication plan includes the context of the announcement to explain the reason for the change to the incumbent personnel and to the client. Also, it includes the service provider's scripting and message for the town hall meetings.
6. Do a site Q&A; revisit.
   - Address gaps from the discovery planning meetings.
7. Sign the contract.
   - The service provider contract needs to be signed so it can be executed, including all the commercial terms agreed to. This allows the setup of the contract in the service provider's system for go-live dates.
8. The owner announces to the incumbent management the decision to part ways, and creates an in-plant formal communication stating the reasons for the change.
   - If no major obstacles are identified, the site communicates the upcoming change to the incumbent and to plant personnel per the communication plan.
9. Hold town hall communication meetings.
   - The service provider will have a Communication/Rollout Team to lead the meetings. The meetings are designed to provide employees with information concerning the new service provider and the scope of change. These could be held off-site at a hotel or other meeting area.
10. Begin to take applications.
    - The new service provider will take applications and make job offers continually until the go-live date (using mobile clinics to speed the drug screen process and reserving meeting rooms in a nearby hotel to take applications if no hiring office is nearby).
11. Identify site scopes of work.
    - Site scopes of work are identified for ongoing and incomplete work. They are especially focused on work that will be incomplete or in progress on the go-live date.

A complete transition spreadsheet tool is provided at www.maintenanceinsanity.com for your use. It includes detailed questions for each functional group to be resolved to prevent items from falling through the cracks during this transition period. A separate timeline is provided as a "double check" for the facilitator to compare his or her observed progress with that of the functional owners' combined efforts shown in their weekly readiness scores. The new service provider will furnish the site this transition performance evaluation in the form of a transition scorecard. The readiness scorecard will monitor the service provider's performance of the transition and will be reviewed by management representatives on a predetermined schedule.

A complete generic milestone plan is shared at www.maintenanceinsanity.com to show the projected timeline to be accomplished. Figure 16.2 gives a portion of a milestone plan.

New Service Provider Transition Kickoff

Here is a list of items to consider for the new service provider's transition kickoff with the site to replace the incumbent:

1. The site must provide its present worker assignments for all crafts and functions to allow you to understand your resource requirements. Will you be taking over any pack, load, and ship function that is directly linked to operations and cannot be interrupted without creating missed orders because of their sold-out status?
2. The site will provide any available training records, but you may need to get the rollover employees to supply their own copies (crane operator, forklift, welding certifications, DOT certifications, track mobile, etc.). You will need to request the customer requirements for a background period (7 or 10 years) and any exemptions the owner may wave (GED requirements, etc.).
3. Service provider training verifications will have to occur prior to the go-live date to prevent any disruptions. The site must agree to support an early signing of the contract to allow some overlap time for the service provider to do training and testing and processing prior to go-live date. Is there a transition budget to cover any required training?
4. The notification timeline for the incumbent employees will be decided. You will need to be available as soon as the announcement is made to help reassure the employees about their future employment. Set up times when the spouses can also be involved for sharing company information. Explain to site personnel any concerns over any benefit differences with the new service provider compared with the incumbent.

TASK DESCRIPTION	ASSIGNED TO		PRIORITY	DUE DATE	= Overall %	STATUS	COMMENTS
	Owner	Provider			% COMPLETE		
Create the IT Survey Worksheet.			2: Prior to moving onto site				
Determine site connectivity to outside Service Provider offices from plant site offices.			3: Between moving & working on site				
Determine and document computer needs for jobsite.			2: Prior to moving onto site				
Drop-Ship Customer-Complex computers, networking equipment, and printers.			3: Between moving & working on site				
Secure database for scaffold tracking.			3: Between moving & working on site				
Set up and test computers & networking equipment.			3: Between moving & working on site				
Evaluate phone requirements and procure phone system & phone line(s).			3: Between moving & working on site				
Procure fax line(s) for jobsite facilities.			3: Between moving & working on site				

FIGURE 16.2 Milestone plan

5. You will need to establish times for day and shift employees to attend the town hall sessions to receive their information packages and applications. They will be told what to bring and assigned an appointment for processing. Make it clear that you will not be reducing head count or wages as part of the initial rollover.
6. Can you obtain a mobile medical unit that you can use for the employee physicals and screening tests? Is there an on-site area available to use for processing? Are new service provider hard hats and boots the only PPE that you have to provide?
7. Are there any site activities (family day, etc.) that you will be allowed to attend? This would show a true partnership.
8. The new service provider will have more flexibility in the workforce to adjust up and down to demand since everyone will be under one company and you can cross-train to grow your people.
9. The site is expected to give the new service provider a first shot at all the new work in the future. You will provide the multicraft skilled employees. It will be up to the service provider to tell them to go to another contractor if you the service provider cannot handle their requests.
10. You must have an open relationship. You should have interactive safety meetings with pictures and sketches to make points and ensure understanding of all incidents.
11. You will need to verify all work schedules and any desired changes by the site.
12. The site will be allowed to see résumés and have interviews with your selected key personnel to ensure cultural fit. You will need to get their buy-in to ensure an understanding of your organizational chart as soon as possible. You will also need the org chart for the owner's site personnel.
13. All the incumbent's signs and posted information must be fully replaced by the new service provider's materials.
14. Connectivity needs to be verified with the service provider systems and any site-provided computers, planning and scheduling work processes and integration with all crafts, interface with the site computer systems, etc.
15. Once all critical positions are identified and the rollover is known, backfilling will be a top priority.
16. You will use this transitional spreadsheet as your starting point for your discussions to gain alignment for the service provider and site partnership.

17. For rollover employees, you will need to request pay stubs and start date with the site from the incumbent employees that apply. This is to verify insurance coverage, vacation eligibility, and wage rates. The site plant manager will be assured of your fair treatment of the incumbent employees during the transition. Be ready to address their concerns and ease their fears by being available for discussions as soon as they are notified on-site that the incumbent has lost the contract.
18. The Transition Team members will be on-site off and on through go-live to train and ensure understanding of their processes by admin and operations. They will work with site personnel to resolve open issues and to finalize data for setting up the accounting system communication. You will need to address the IT requirements for invoicing also.
19. The IT people need to verify the system interface and connectivity and let you know how you gain access as soon as you get on-site. They will work together to create employee site IDs and build lists for classifications and pay rates. No work is to be done without a charge number. Provide the site with an early warning process for flagging potential overruns. An initial priority is getting your people paid without interruption.
20. You need to get verification on the required criminal background checks and also verify if any grandfathered exemptions will be given by the site. Obtain employee site rosters to track commitment to roll over to you. Develop communication flyers to share with potential employees. Hand them out during the meet-and-greet session and then schedule times for individuals to be processed.
21. A local recruiter can help backfill any openings. Identify all testing (weld certification, crane operator, etc.) that must be done prior to go-live and others (fit test, etc.) that can be done post hiring. If there are any medical "holds," you will have to backfill until the person can get treatment and be retested. Request the site shift schedules so you will know when to hold sessions to reach all the rotating shift employees.
22. You will need to work with the site to develop a demobilization plan for the incumbent and a mob plan for the new service provider to ensure a smooth handoff of facilities and workload. You will need the site engineering specs (welding, painting, insulation, etc.) that you are expected to meet. Identify the owner's third-party inspection company to start developing a relation-

ship. For QA/QC, you will supply the service provider with a quality manual for use at the site.

23. You will need to validate craft needs versus the people available. Once the contract is signed, get it to the key players to ensure understanding for implementation. The service provider site manager is to write a summary of the contract for sharing.
24. The service provider's org chart and site structure have to be finalized for sharing with the client. Your communication plan must include who is to be at each meet and greet and then at each processing session for rollover and new employees. You will need a contingency plan for key positions that do not roll over. Identify quick transfers for temporary service until locals can be obtained. Processing these people into the sites must be part of the plan. You must weed out any undesirables and replace them with new hires.
25. A site safety manual will need to be produced for the service provider's programs.
26. For go-live, you will need to concentrate on getting the people transitioned and putting critical support processes in place. You will have to maintain the status quo for about 90 days to develop your baseline before initiating your improvement processes. You must track performance during this period to be able to show the impact of completed improvements. You will also have to identify the needed training for your people.
27. You will need to let potential rollover employees know that this is a long-term contract with extensions based on performance with the intention of being a true partnership with the site. Ensure those employees that no lapses will occur in the incumbent employee insurance coverage and benefits.
28. The service provider will add value to the site and to the rollover employees through training and process improvement efforts once the transition is complete.
29. Safety will be your top priority during this stressful period, and you need all of the employees' help to be successful.
30. The site manager needs to kick off the sharing sessions with the incumbent employees, telling why the company decided to make the contract change and explaining that the company selected the new service provider to help move the site to the next level of performance.

The next list is based on numerous field applications of this process and will benefit you by preventing the duplication of errors made by those who came before you. To leverage the advantage of experience, the following lessons learned are offered to benefit your transition:

Transition Lessons Learned

1. The new service provider's site team should be built with a base for success for future performance. Bring key positions on board as early as possible to get involved with client meetings, daily reporting and site practices, and required training and to meet the site people. Review the proposed selections with the client.
2. Site communication must be clear and give dates for the full transition to the new service provider. This is critical to easing tensions and concerns. We suggest the client publish a weekly newsletter to minimize rumors and answer questions. People worry during change.
3. Both the work tracking software and management software must be implemented from the start for all services.
4. Both early identification of equipment needs and expedited deliveries are critical to a smooth start-up. This includes computers and the required access and connections to function. Determining a mutually agreed-upon disposition plan for the incumbent's site equipment is very beneficial to the site and contractors. Vehicles are key to moving the people for job performance. Lead time is required to obtain vehicle passes, to select equipment staging areas, and to clear site entry processes. Document the site requirements to minimize changes to the initial orders.
5. Identification of key partner contacts for the new service provider and client will allow relationship development by each functional group during discovery and start-up.
6. The client's site organizational charts and site plot plans are required reference tools for discovery.
7. On-site transition offices need to be available early and throughout all phases of transition.
8. A demobilization strategy must be required and shared by the incumbent contractor to allow access for the new service provider's mobilization to the site. Site real estate is always limited.

9. A site work flow process must be understood as a basis for all business activities (request of work, planning and scheduling, estimating, invoicing, auditing, execution, etc.).
10. Contract reviews are needed to clarify and resolve issues to gain understanding by all parties.
11. Friday and Saturday processing sessions should be scheduled to ensure that the personnel transitional process is performed with minimum disruption and positive impact on cultural change. Hold early off-site meetings with potential employees and coordinate with the local safety council for weekend support. Incumbent rollovers have a big advantage over new hires.
12. You can drive standardization of site practices while incorporating identified improvements. This is your opportunity to do it the way you have always wanted it done.
13. Safety awareness and field presence need to be enhanced during the start of the work phases.
14. A phased implementation may be best when it is based upon the number of employees, size and layout of the site, impact on operations, etc. This is a joint decision.
15. The equipment rental agreement needs to be clarified and set for hand-over items from the incumbent for a seamless transition.
16. Contingency plans should be prepared to address unresolved issues as they occur. Document and share these agreements with all parties.
17. Transition work orders and purchase orders need to be set up to test formats and process flow early in the transition to debug all systems.
18. A true joint effort is required for asbestos transition to meet all regulatory requirements and permit needs.
19. It is vital to address immediate needs (emergency preparedness and response plans, safety expectations, etc.).
20. Stewardship (KPI) packages are needed for short- and long-term measures for safety and performance to drive a continual improvement process.
21. Transition progress is tracked by scorecard activities and scheduled functional milestones to compare percent completions for confirmation of progress (two touchpoints).
22. Initial work lists are developed for start-up weeks based on site area visits to verify the head-count hiring requirements when the transition is in phases.

Even when equipped with a good transition plan, you must be prepared to handle the unexpected. This final list gives a brief contingency plan example:

1. In the event that the current provider decides to remove all assets and personnel from the site earlier than the set date, the new service provider must have a "rapid response" plan preapproved with the client for all its services.
2. Operations' critical tasks would be covered by temporarily transferring the new service provider's employees from various regional locations until full-time resources can be obtained. Evening and 24-hour coverage would be supplied as required.
3. The client site personnel will transfer the work order backlog for all trades to the new service provider's planners with as much advanced notice as possible prior to the start date to allow estimate sheets and job plans to be ready.
4. The new service provider's regional resources would allow the service provider time to recruit locally and roll over any interested site-experienced employees. A minimum of seven days with one weekend for processing would be required to notify the employees needed and complete the roll-over process.
5. The new service provider must procure additional equipment and materials for its crews during the time the material is being obtained.
6. All temporary regional support would be phased out as the local workforce is hired and cleared for work.

Following these examples will ensure a smooth transition with minimum disruptions of services by all parties involved. It will allow you to compare existing and new processes to eliminate any new maintenance insanity infections from taking root and spreading during the period of change.

The next chapter will share successful processes for managing a turnaround for your facility. Turnarounds come with their own form of maintenance insanity because they always include discovery work that was unexpected and requires out-of-the-box responses.

CHAPTER 17

Shutdown and Turnaround Solutions

Just like with maintenance insanity, life must be managed to achieve a cure. This is especially true for an industrial turnaround (T/A) situation. You have a large defined amount of work that must be achieved in a short period of time (to minimize lost revenue from missed production). The insanity picks up when the discovery work shows up. The same old way of doing things does not work in this high-pressure environment. Take charge and resolve each issue to meet your start-up target.

The following explanation will help your organization to improve shutdown planning, scheduling, and execution processes for the work you perform and for the actions you outsource. Do an initial check step to test your perceptions versus reality. Do your observations for past turnarounds reveal that:

- For many add-on jobs, work outgrew resources?
- Work definition and coordination were poor?
- The outage start date was moved several times?
- Parts and materials were not available due to changes?
- There were delays due to lack of safety preparations?
- Start-up delays occurred returning to service?
- Work was planned as part of the daily workload without dedicated resources?

Versus

- We are not *that bad*. In fact, *we are OK!* (maintenance insanity)

To get you started, Figure 17.1 gives you a Six Sigma process map for a generic shutdown process.

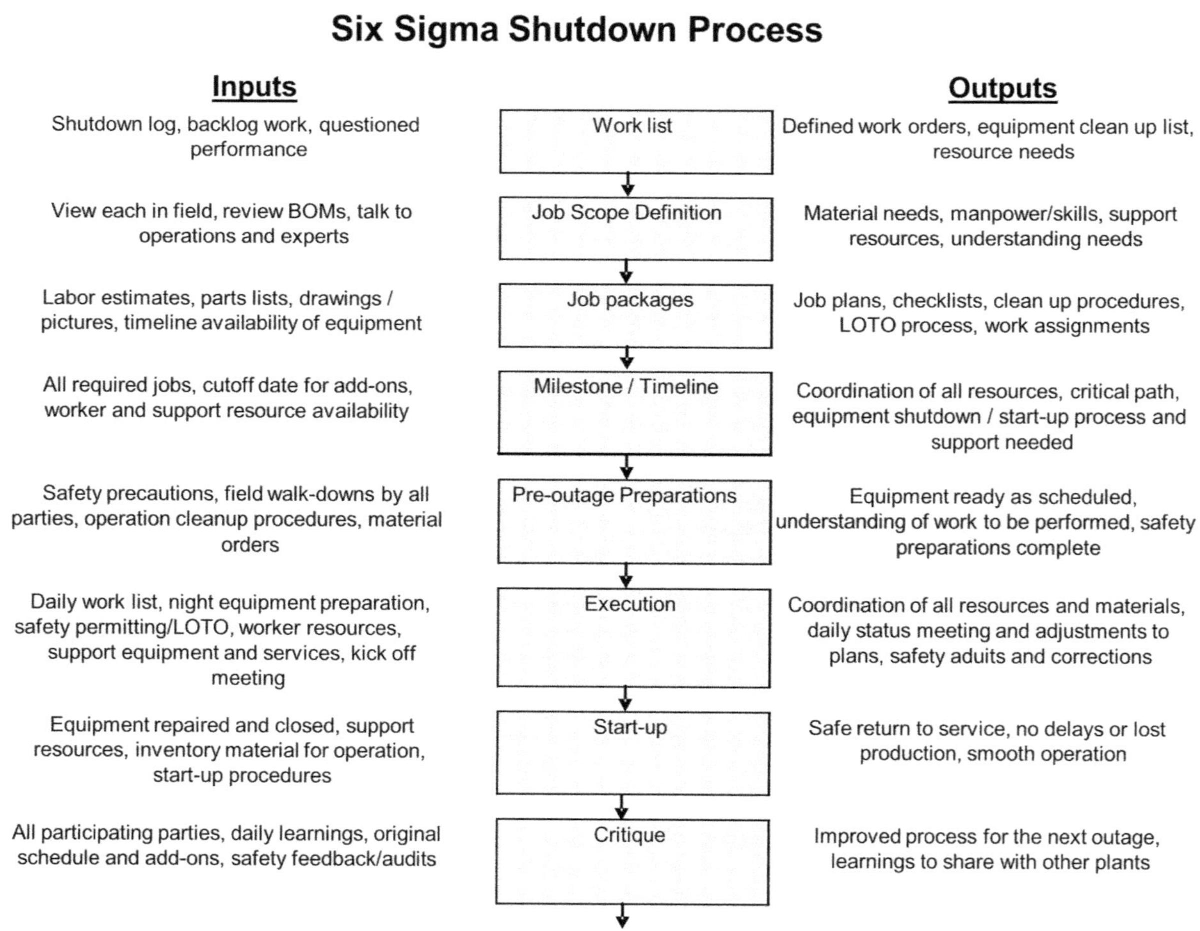

FIGURE 17.1 Shutdown process map

Use this example as a starting point for your turnaround process map to understand each step.

> To have a successful shutdown process, you must make a change to get a change. Assign dedicated shutdown roles, set and enforce a cutoff date for your common work list, hold to an established start date, and create a single master work list and schedule. The shutdown planning and scheduling guidelines must include the planner(s) assigned full time to manage the master work list with adequate lead time. A good rule of thumb is one-month preparation time for each week or partial week of outage scheduled with the cutoff date for adding new work set at a minimum of three weeks prior to the start date. All parties must be involved in field walk-downs to understand the work scope and execution.

Your turnaround vision should be committed to operating safe, productive, and environmentally sound manufacturing units. This can only be accomplished by ensuring that all maintenance turnarounds are performed on an established routine basis. These turnarounds will be done in the safest, most environmentally friendly, most cost-efficient way, with the shortest duration possible. Minimal resources, both internal and external, will be used in a manner that will limit overtime and reduce stress on turnaround participants. All of this will be focused on making these turnarounds transparent to the owner, your customers, and the contractors.

To assist operations and maintenance in determining turnaround interaction windows, a yearly turnaround calendar must be maintained. This calendar shows all turnarounds, plant-wide, and the date, duration, and area in which turnarounds are scheduled. The calendar must be able to be viewed online. To submit a turnaround, communicate the required information to your operations/maintenance representative. Conflicts that may develop between turnarounds will be resolved by the maintenance and operating areas involved in the conflict and will be facilitated by the Site Leadership Team for management approval.

Executing a shutdown in a safe and cost-effective manner, and doing it utilizing minimal resources both internal and external, requires cooperation among all the parties involved. Operations, maintenance, and contractors must work together to ensure the success of the shutdown. The shutdown process should include work scope development, planning and scheduling, execution, and follow-up for expense and capital work.

A shutdown log should be created to capture all the jobs that require a shutdown to complete. With SAP, the work order backlog would be available to be accessed and updated by operations and maintenance and engineering during the year. Jobs discussed in the daily planning meetings that are determined to be shutdown items will be added to the backlog.

Work scope development is the process for determining what work, from the shutdown log, is to be done during the shutdown and creating the specifications for that work. This is a cooperative process involving operations and maintenance with capital project input. One of the major considerations during the work scope development phase is determining if a job is actually turnaround work or if it is work that could be done during normal plant operation. The maintenance shutdown planner and the production assistant for the appropriate area are the key players for shutdown coordination. A cutoff date must be established and enforced for freezing the scope of the work to be done. Allowing maintenance insanity to rule a turnaround will result in a longer list of work than could ever be completed in the set duration of the shutdown.

Proper planning and scheduling is essential to the success of the shutdown. There are several complex steps that must take place when planning a shutdown. To ensure all the steps are considered, the planner should fill out a checklist. A typical example of a master shutdown checklist is provided at www.maintenanceinsanity.com for you to consider when initiating your T/A planning.

Detailed job plans should be created for each job on the shutdown plan. These plans should include a general job description, crew size, estimated duration, safety tips, permits required, materials needed, tools needed, and detailed work instructions. Since a large number of jobs are repeated each year, task lists or PM/stored plans should be created in SAP or another computer system.

With limited resources available, scheduling becomes a critical part of the shutdown. The planner should work with operations, the maintenance supervisor, and the contractors to ensure that all work is performed in the most efficient manner. A shutdown schedule should be created prior to the start of the shutdown showing which jobs are to be done each day by which groups. The planners should ensure that all the materials needed for the shutdown are on the jobsite before the start of the shutdown.

Cooperation between operations and maintenance and engineering is essential during the execution phase of the shutdown. Daily meetings should be held to keep everyone informed on the status of work and of any problems that might exist. The following list gives the roles and responsibilities for the required players for a major turnaround:

Shutdown Roles and Responsibilities

- Shutdown coordinator (key role from operations)
- Shutdown planner/scheduler (key role from maintenance)
- Area maintenance manager
- Maintenance team manager
- Area maintenance coordinator
- Contractor coordinator
- Material coordinator/area stores representative
- Reliability department representatives
- Mechanics
- Production supervisors/building coordinators
- Flex/contractors and support resource representatives
- Engineering representatives with management support
- Area safety representatives

A daily shutdown status meeting must be held midafternoon or prior to shift change each day to review the following items:

- Check for jobs completed today (update the master schedule).
- Learn from delays/problems to adjust the schedule.
- Determine carry-overs and their impact on tomorrow.
- Go over tomorrow's jobs: equipment preparation, resource conflicts, permitting, and material staging.
- Make action item assignments and follow up daily.

Focus on the basics to make sure the following are correct: work scope definitions, the operations plan for equipment and systems shutdown and start-up, mechanic job packages, resource staffing and coordination, and safe execution. You must also track progress through standard shutdown measures like turnaround duration, turnaround labor-hours, turnaround costs, scope change, percentage of work added after the cutoff date, start date compliance, number of times the start date was changed, and execution performance field and admin audits.

Always hold a shutdown critique to drive continual improvement. Capture what was done well and what can be improved. With all parties involved in the outage, review activities for each group by the following stages: planning and scheduling, pre-turnaround preparations, work execution, operational start-up, and post-shutdown activities. To ensure that your turnaround processes will be successful, have a true team effort from all involved parties, visibly demonstrate management support, set expectations with accountability for sticking to the schedule, and celebrate success and share learnings with others. What gets rewarded is what gets done.

Visit www.maintenanceinsanity.com for a complete turnaround guide and for forms like a master turnaround checklist to review the list of tasks to be considered for each shutdown.

When applicable due to size or duration, a Turnaround Planning Team will be established when an operating or maintenance area agrees that the complexity of the outage requires special attention. The Site Leadership Team members for the site will set a strategic planning meeting with the functional leaders of the plant to be shut down and the maintenance manager. An overall shutdown coordinator will be named to lead the team (coordinator, operations coordinator, planner, and maintenance supervisor).

The team will develop the strategic turnaround plan. This plan will include, but not be limited to, the following:

1. Define the turnaround purpose and the performance goals.
2. Identify strengths.
3. Recognize external and internal influences on turnaround performance
4. Identify weaknesses that need to be addressed.
5. Assign/accept responsibilities to plan and execute the work.
6. Initiate systematic programs to cost-effectively and efficiently achieve turnaround goals.

Once developed, the strategic turnaround plan will be formalized and issued to all major participants in the turnaround and will become part of the larger turnaround plan.

As required, temporary membership of the Turnaround Planning Team may also include representatives from stores, purchasing, safety, instrument tech, environmental, contractors, and engineering, as well as the contractor coordinator and any other organization so deemed by the shutdown coordinator.

The Turnaround Planning Team will also set overall tentative milestones for the turnaround. These milestones should include dates for beginning planning, issuance of cost estimates, ordering of special materials and spare parts, unit turnaround date, beginning work date, welding window dates, turnaround completion date, and cleanup completion date.

Before a turnaround can be planned effectively, the objectives of the turnaround must be established and conveyed to the turnaround's key participants. The turnaround management plan (TMP) will document the details of all elements that are involved in planning, staffing, execution, and completion of a turnaround, and therefore the plan defines what the turnaround intends to accomplish. The key players include

- **The planner.** This person is responsible for maintaining the turnaround calendar and initiating the turnaround planning phases of the turnaround planning process. The planner is responsible for managing the work scope development and planning/scheduling phases of planning the turnaround including interfaces with all capital project activities.
- **The production assistant.** The PA is responsible for managing the equipment preparation for the work execution phase of the turnaround and reporting to operations the daily status of the turnaround.
- **The maintenance supervisor.** This person is responsible for the work execution phase of the maintenance activities by the owner and contractor mechanics. The maintenance supervisor will manage blocks of work during a turnaround. This supervisor is ultimately responsible for ensuring that the turnaround work is completed, by schedule, in the most cost-effective manner possible.
- **Site safety.** The job of this person is to support safety preparations and audit activities during the outage.
- **Shutdown coordinator.** This person has overall responsibility for managing the total outage including all capital and maintenance activities. He or she is the interface between engineering, operations, and maintenance to optimize work flow and interactions of all the groups. The shutdown coordinator drives the up-front alignment of equipment shutdown and cleanup to have the appropriate items ready for the scheduled capital or maintenance planned jobs. This coordinator is responsible for the critique and follow-up phase of the turnaround.

For work scope development, the operating department maintains a log of all work that can only be done during turnarounds. The log entries begin when a turnaround ends and continue until the add-on cutoff date of the next turnaround. At the beginning of the turnaround planning process, operations creates a work list from the log entries. This work list is sent to the PA and planner to begin planning.

Work scope development is the process for determining what work, from the work list, is to be done during the turnaround and creating the specifications for that work. This is a cooperative process involving operations, maintenance, and engineering. One of the major considerations during the work scope development phase is determining if a job is actually turnaround work or if it is work that could be done during normal plant operation. If it is determined that a job is work that can be done during normal operation, the work will be added to the normal backlog or scheduled as appropriate.

During the work scope development phase of the planning process, operations will assign an operator to work with the planner. This team will set a cutoff date for all new work to be added to the work list, will begin reviewing the turnaround log, and will begin developing the turnaround work scope. Any work received after the cutoff date will be considered add-on work and will be accounted for separately. These costs will be shown as additional costs to the turnaround. For normal turnarounds, the cutoff date should be approximately six weeks before the turnaround date. The time from the cutoff date to the beginning of the turnaround is the scheduling and cost estimate phase of the planning process. During the scheduling phase, the timeline chart will be developed, and the required resources will be applied to the activities.

To ensure accuracy between what operations needs and maintenance plans, a job scope sheet will be completed on all large major jobs done on the turnaround.

Planning and scheduling steps include:

Turnaround Procedures and Stored Plans

During the planning and scheduling phase, it is the planner's responsibility to ensure that all turnaround processes and maintenance work have procedures or stored plans, as necessary. The planner will contact all participating groups to establish that documentation is available. These groups include operations, maintenance, contractors, and engineering. These procedures and stored plans should provide the employees doing the work all the information required to perform the activities in the most efficient and effective manner possible.

Master Turnaround Checklist

There are several complex steps that must take place when planning a turnaround. To ensure all steps are considered, the planner must fill out a master turnaround checklist. This checklist provides for milestones to be assigned to each activity applicable to the turnaround being planned. As an activity is completed, the completion date will be documented on the checklist. At the completion of the turnaround, the completeness and timeliness of the checklist will be measured. A graph will be maintained by the planner showing these measures for all turnarounds. During the turnaround planner's critique meeting, which follows each turnaround, the planner will review the master turnaround checklist to determine if new activities need to be added. See a sample checklist at www.maintenanceinsanity.com.

Walk-Throughs

There are several "walk-throughs" that must take place to ensure that all aspects of a job have been identified and that this information is presented to the supervisors and mechanics in a manner that optimizes job performance. Each job must have a field "walk-through."

Operations/Planner Walk-Through

The first and most basic walk-through is where the operations representative shows all the jobs to the planner. This walk-through is conducted after the work list has been issued and the data has been input into the turnaround work list. Once the work list is entered into the database, walk-through information notes are made during fact-gathering walk-throughs.

Relief Device Information Walk-Through

Another very important walk-through, which is conducted shortly after the issuance of the work list, is the relief devices review walk-through. All relief devices that will be removed during the turnaround will be reviewed at this time. Digital pictures can be used to show locations for tagging and removal.

The preparation for this turnaround is similar to the preparation for the operations/planner walk-through. A database should be created that contains all the relief devices that will be removed. Unlike the turnaround work list database, the relief device database is not a transient database. The information obtained during the walk-through will be pertinent for all preceding turnarounds. Since not all relief devices are removed during each turnaround, the database will continue to build with each turnaround until all relief devices are included.

Contractor/Vendor/Planner Walk-Throughs

To ensure that each contractor and vendor that will be working the turnaround fully understands the requirements and expectations during the turnaround, the planner should review each job during a walk-through. If the contractor or vendor is performing work for several groups, such as maintenance, environmental, and engineering, or in conjunction with other contractors or vendors, all parties should be included in the walk-through. This will ensure continuity of the contractor/vendor work and will reduce the possibility of congestion, conflicts, and delays.

Upon completion of the walk-throughs, each contractor and vendor will provide the planner with a cost estimate. This should be provided approximately one week prior to the planner issuing the total turnaround cost estimate. The planner will provide this date to the contractors and vendors as part of the walk-through.

Maintenance Supervisor Walk-Throughs

Supervisors assigned to the turnaround should be shown their work during a walk-through with the planner most familiar with their work. If possible, these should be individual walk-throughs by work packages. All supervisors responsible for the completion of a specific work package should attend the walk-through. The walk-throughs should be conducted two weeks prior to the turnaround, minimal. The shutdown coordinator should be involved in the walk-throughs as questions occur.

• • •

Staging of equipment and materials is the next critical step. This process defines a strategy for the control of parking, equipment staging, and material laydown.

Personal, company, contractor, and vendor vehicles will be parked in designated areas, while strategic areas around the operating unit will be designated no parking. The shutdown coordinator will establish these areas along with entrance and egress lanes (safety lanes) for emergency situations. Once these areas and lanes have been designated, the information will be indicated on a plot plan showing all affected areas, in both the operating unit and adjacent areas. The plot plan will then be given to the planner responsible for the turnaround to be circulated among all groups that will be working on the turnaround; this includes the capital and maintenance groups.

As the plot plan is circulated, each group will mark it with requested locations for equipment setup and materials storage. The placement of equipment and materials will be prioritized by the importance of the work being done. Mobile cranes should be placed first, vital part laydown next, then hydroblasting equipment, etc. Once all the groups have put their requested locations on the plot plan, the planner will collect the plot plan and issue it to all groups involved in the outage.

A major focus for all work activities is safety. Turnaround work should be approached no differently than is routine work. Every precaution should be taken to ensure that all participants understand all safety aspects of the work being done. Although turnaround duration is of the utmost importance, it should never overshadow safety. Consider creating a supervisor expectation sheet. Additional operations resources will be supplied as appropriate to prevent delay in job preparations.

Safety Expectations for Turnarounds

It has become a standard practice for maintenance personnel from all plant areas to work on major turnarounds. To ensure continuity of safety expectations, the following guidelines are provided:

Supervisor/Mechanic Responsibilities

1. Toolbox safety meetings
   a. Prior to the start of every shift, crews will hold a toolbox safety meeting.
2. Job safety reviews
   a. Ensure that job safety reviews are conducted before each job.
   b. Verify that job safety reviews identify and address hazards before work starts.
   c. Ensure that all mechanics on the job participate in job safety.
3. Permits and lockout tagout
   a. Verify that equipment has been properly prepared for work.
   b. Verify that the lockout tagout process has been properly executed.
   c. Ensure that the information on the permit(s) is discussed with all mechanics working on the job.
   d. Ensure that the mechanics sign the permit and logbooks as appropriate.
   e. Sign permits as required.
4. Tools and equipment
   a. Ensure that the correct tools and equipment are being used on the job and that the tools are in good condition.
   b. Ensure that all the correct safety equipment needed is at the job and is in good condition.
   c. Ensure that all mechanics working on the job have and are using correct personal protective equipment.

In order to meaningfully track and update the progress of a turnaround, it is necessary to record job starting times, crew size, actual job duration, and completion times.

Capturing the above data is the responsibility of the turnaround supervisors; this includes operations, contractors, engineering (capital projects), and maintenance.

Cost Management and Forecasting

When determining accurate turnaround cost, all costs associated with the maintenance work and loss of production must be included. Reducing total turnaround costs is of the utmost importance. While there are several ways for reducing overall cost, the three major methods are controlling total turnaround duration, doing innovative job planning, and culling non-turnaround work.

Reduction of Turnaround Duration

Depending on the margins and profitability of the operating plant, it may be advantageous to optimize the duration of the turnaround. If margins and profitability are high, it may be cost-effective to reduce turnaround duration by increasing staffing. This can be accomplished by either increasing the number of employees working on the turnaround or instituting planned overtime. If margins and profitability are low, it may be advantageous to extend the turnaround duration, thus being able to reduce staffing.

Innovative Job Planning

Another method of reducing total turnaround cost is planning work to ensure that it is being done in the most effective manner. This includes new methods, new suppliers, and new materials. It is the planner's responsibility to review all jobs with this concept in mind. Savings created by innovative job plans will be tracked and a measure maintained.

Culling of Non-turnaround Work

On some occasions, work will be included on the work list that could be done outside the turnaround window. This can be work such as repairs on spare equipment and equipment that can be put out of service without affecting unit production. It is the planner's responsibility to review all work and ensure that only work that requires a unit turnaround is included on the work list. If a non-turnaround job is found on the work list, it should be brought to the attention of operations for removal from the work list. If operations is reluctant to remove the job from the work list, the planner should notify the shutdown coordinator. The shutdown coordinator will review the job with the maintenance manager and production manager for resolution. Savings created by eliminating jobs will be tracked and a measure maintained by the planner. A savings form should be created and used to calculate savings for innovative job planning and for culled jobs.

Cost Estimate

Two weeks after the work list is finalized, the planner will issue a tentative turnaround cost estimate. This estimate will include all maintenance work to be done during the turnaround including maintenance, support groups, I&E, and contractors (capital and maintenance). The turnaround cost estimate will be issued to the production manager and the maintenance manager.

Cost Tracking and Control

It is imperative that turnaround costs be controlled throughout the total duration. This includes T/A development, work execution, and the post-work period. This is accomplished by use of a set of reports and a database. The reports are the daily report and the category totals report. The database from which these reports are created is the daily costs database/SAP or other tracking software.

Daily the shutdown coordinator will review the reports for cost anomalies. If an expected cost rise is found, the coordinator will thoroughly investigate the cause of the cost. The daily reviews will begin approximately six weeks prior to large complex T/As and will continue as long as all open POs are valid, which is for approximately 8 to 10 weeks after the T/A.

Contractor Management and Controls

The contract coordinator will get a representative assigned for capital and expense jobs to be a liaison between the contractors working on the turnaround, both in-house and outside, and maintenance. The specific responsibilities of the contract coordinator include

- **Establishing and communicating job-specific expectations.** The contract coordinator has to develop generic expectations that are valid for most large critical jobs involving contractors. The expectations of each contractor involved in large critical work will be specific and comprehensive and will include the specific objectives of that contractor, schedule information, the items the contractor is responsible for, reporting requirements, applicable procedures, etc.

- **Performing.** The administrative requirements of procuring and processing. The contract coordinator and the representatives will work together to secure needed passes, issue purchase orders, etc., that are required. In addition, the contract coordinator will verify and approve payment for the contractor services.
- **Ensuring communication.** The contract coordinator will ensure that communication to and from the involved contractors is maintained throughout the planning (cost and resource estimates, schedules, prejob meetings, etc.), execution (status meetings, customer communication, written status reports/data, etc.), and post-completion (critique, etc.) phases.
- **Auditing.** The contract coordinator will periodically audit the contractors' work to ensure that the objectives are being met and that the safety and quality are acceptable.
- **Measuring.** The contract coordinator will, with input from applicable sources, complete and report on the overall performance of the involved contractors.
- **Tracking improvements.** The contract coordinator will be responsible for documenting, communicating savings, and following up on any contractor-related improvements that were identified.

To tie all the loose ends together, the final process is to capture the learnings from each turnaround.

Turnaround Critique Meetings

To ensure open communications from all parties involved in a major turnaround, it has been determined that three turnaround critiques are required. The first two critique meetings are informal and should be called within the first week after the completion of the turnaround. These meetings are the planner critique meeting and the supervisor critique meeting. The third meeting is the turnaround critique meeting and is the formal critique of the overall turnaround.

The purpose of the turnaround critique meeting is to review the overall turnaround for items that were done well and those that need improvement, with the primary focus on anything that will improve the turnaround management process.

The agenda for the turnaround critique meeting will be as follows:

- Planning and scheduling activities
  - Maintenance
  - Operations
  - I&E
  - Contractors
  - Engineering and capital projects
- Pre-turnaround activities
  - Maintenance
  - Operations
  - I&E
  - Contractors
  - Engineering and capital projects
  - Operational turnaround
- Work execution
  - Operations
  - I&E
  - Contractors
  - Maintenance
  - Capital projects
- Operational startup
- Post-turnaround activities

Within two weeks of the completion of the turnaround critique meeting, the shutdown coordinator will issue a critique memorandum. This memorandum will follow the structure of the meeting and will include all action items, with responsibilities and completion dates. Action items from all turnaround critique meetings will be maintained in a database. This database will be reviewed and updated on a quarterly basis. A report will be issued showing two views of the action items. One view is all action items, completed and in process, by turnarounds, and the other view is all in-process action items by area or person responsible.

Action Item Follow-Up Meetings

If an action item remains on the turnaround critique memorandum for one month beyond its completion date, the maintenance manager will call a meeting with the concerned parties to finalize an action plan for the completion of the action item and will establish a "must-be" completed date.

Turnaround Summary Memorandum

Within two weeks of the issuance of the turnaround critique memorandum, the cost analysis will issue a summary of the turnaround. This will allow for the settling of SAP costs and labor-hours.

The Turnaround Blinding and Isolation Solution: Outsourcing

I will end this T/A process with an opportunity to consider a new strategy. Finding and using existing blinds from a plant is a very good example of maintenance insanity. The problem of managing blinds for all the connections that have to be tested during any turnaround is complicated by all the different work groups involved in a confined area working on a compressed work schedule. Who owns this function? The easy answer is "no one" or "everyone," but in either of these cases the result is never optimum. There is too much loss, waste, and discarded materials. You end up spending excess time looking for and handling each piece. A lot of the old blinds that you find are not in good enough shape to be used or cannot be refurbished (additional cost).

I am sure that your system for managing blinds has been looked into several times. The conclusion is usually that we do not have a perfect system and there is a recognized need for better blind organization and management. The problem is that most plants have multiple supplies of all sizes of blinds squirrelled away in every shop and maintenance area across their facility. Your general contractor may even have a separate supply of test blinds for new construction. There are also mixed signals from management about letting go of ownership and trusting outsourcing to handle this task. So far, your current system appears to be working, but organizationally speaking, it could be better even though you seem to be getting it done.

You have tried to get stores or some other support group to manage this function for maintenance and construction, but stores would not accept it. When everyone owns a system, no one really owns or manages it. Maintenance is busy managing the work, so keeping track of blinds is not high on its priority.

Then there is the problem of having space or a designated area for year-round blind storage including manning the checking in and out of the blinds during turnarounds. When you stay open to all options, the bottom line comes down to money spent versus money saved. Economics and safe quality service must be addressed by the final solution.

An outsourcing partner can offer a service for any-size turnaround; it is capable of showing up for a large outage in the correct-size trailer to handle the requested inventory (soft-side 18-wheeler with fork truck access to a smaller trailer with access for removal and restock) and then operating the location like a blinds storeroom for the entire outage and each major capital project. As the owner of the blinds, the outsource company manages all blinds and racks to track who checks them out and returns material for each job. The outsource company tracks by the sign-in and sign-out sheets for all users to ensure accountability. Its on-site people should have turnaround experience to be familiar with the plant environment and pace of turnaround activities. The pickup location would be staffed to be available 24/7 for the duration of the outage. For very large turnarounds, the outsource company will require a laydown area by its trailer for pallets and management of the total inventory. For smaller needs, it can also ship a single blind or a lockable rack full of blinds to the jobsite, and it is then your responsibility to manage them. Most have shop capability to provide 24/7 service including emergency response for any need. The outsource company should be able to provide same-day shipping nationwide for in-stock items.

The outsource company is renting the blinds to the owner for the turnaround duration. Based on the cost to buy versus rental savings for a major refinery turnaround, savings of up to 38% have been achieved compared with the normal purchases made in the past for the same outage by using the rental service. Always ask for reference contacts for end users of the rental service for turnarounds and projects. See if they are repeat customers of the service to determine the level of their satisfaction.

After a review of your present inventory of blinds and their condition, ask if the outsource/rental company would be willing to negotiate a discount on your rentals for up to the first two years to offset the value represented by any materials the outsource company accepts from your present stock if you want to have materials removed from your site. If you want to keep your inventory, see if the outsource/rental company can supply all the needed materials for any size of outage or project. At the end of the shutdown, the outsource/rental company's on-site personnel working at the checkout area should assist with the inspection of dumpsters to recover and clean up any discarded blinds.

To meet all your needs, make sure that the outsource/rental company can provide hydrotest and isolation blinds, spectacle (figure eights) and paddle blinds, vented paddle blinds, bleed rings, and blind flanges. Determine the range of sizes the company keeps in stock (paddle blinds and blind flanges ranging from ½-inch to 60-inch diameter for 150 psi to 2,500 psi class rating). The company should be available for same-day local or nationwide shipping from inventory stock. It is critical to select a vendor that stocks what you need to have it available immediately. In addition, verify if the company can custom-fabricate additional sizes to meet stringent specifications, unique requests, and urgent turnaround times.

Consider a shop visit to confirm the outsource/rental company's capabilities. Its machine shop should include plasma and water-jet cutting tables capable of cutting carbon steel up to 12 inches thick and stainless steel up to 4 inches thick. Lathes and vertical boring mills should handle up to 72-inch diameter. Check to see if the company is able to offer gasket resurfacing in its shop or in the field for damaged flanges and valves. Also, does it have the capability to cut and machine any size of blind needed with fabrication and then to ship the same day in most cases? The blinds need to be color-coded and stamped for easy identification. Blind flanges must have lifting lugs attached for safety. The smaller blinds should have a hook handle, and the larger ones should have a T-handle for ease of handling. The company should be able to ship blinds individually or on lockable racks that are safe and easy to use. The blinds must be flagged for the unit where they will be installed (on the blind and rack). They will need to be palletized per piece of equipment to maximize efficiency (example T-1). When the customer supplies details ahead of time, the rental company should also be able to identify with the tie-in point or blind list location for ease of installation. This identification would also help with removal from each unit and recovery of as many blinds as possible.

Renting from a blind rental company must be simple and smart: you must save money, eliminate site long-term storage issues, and reduce tax-related inventory concerns. Whether you need one blind or 10,000, the rental company must meet even the most demanding timetables, from a scheduled turnaround to an emergency caused by Mother Nature.

The rental blinds must be manufactured according to API 590 standards and stamped with size and rating. In addition, they are to be color-coded and have handles for easy identification. For example, all 150-psi-class paddle blinds can have a blue handle, and flanges can have a blue rim. The end user needs to have secure, lockable blind racks for easy access, convenient storage, and simple deployment throughout its facility while minimizing the risk of loss or theft. Ensure that you receive new-condition blinds palletized or crated or on a lockable rack. The rental company must make returning your rental order easy too. When you are ready, give the rental people a call and they should schedule a convenient time to pick everything up.

Blind rental services can support refineries, chemical plants, power plants, and terminals in performing facility turnarounds safely, cost-effectively, and efficiently. Your selected rental company should have access to trained turnaround personnel, fabrication shop facilities, and skilled crafts to support your site blind management activities. Get the company's people who will be working on-site to be involved with all pre-outage field walks to help identify and verify where the blinds are to be used. If you need the service, ask if the company has technical qualified technicians to handle your torquing services and other capabilities for future interactions. Your goal should be to create a lifelong partnership based on the company's quality, service, and value. Check the web for blind rental companies in your area.

• • •

Curing maintenance insanity, especially for turnarounds, requires the structure and processes provided in this chapter. The T/A critique will ensure that you capture learnings from this outage and apply them to be better with your next turnaround.

In the final chapter, I want to share a final solution to your maintenance insanity showing you what is possible when you benchmark and use processes implemented to survive some unusual conditions.

CHAPTER 18

A Glimpse at the Future

I want to close the "Maintenance Insanity" Cure by showing you how you can benefit from following the example of others who had to overcome very unusual conditions and circumstances to survive any situation that might occur at their facility. You get the gain without any of the pain if you are open-minded enough to see the potential for your site and give the processes a try. Sometimes your situation dictates your solution, but from it comes an "aha" moment for the rest of us where all the pieces just seem to fit together. As Figure 18.1 suggests, solutions occur when all the pieces come together. The insanity is reduced by the number of solutions we achieve.

From 1996 to 2000, I had the privilege of doing a greenfield construction and start-up of a chemical plant in Asia in the industrial complex of Singapore. Part of the land was still being reclaimed from the ocean when we arrived. This story describes how we developed work processes to do things the way we had always wanted to do them.

Sometimes necessity drives us to take positive actions to address unusual circumstances. Sakra Island, the island on which the chemical plant would be built, was part of the Singapore industrial complex made up of five separate islands that were being combined into one island by reclaiming land from the sea. When we started construction, we knew that our plant would only have ferryboat access until a road could be completed several years in the future. Because of its isolation, it would take first responders and support resources a minimum of four hours to provide any help. Therefore, the personnel on-site for each shift had to be trained to be qualified to safely handle any situation (fix minor failures, patch until help arrived, shut down and isolate, etc.) that might occur.

The following solution defines a very effective maintenance and operating process that was developed to address our immediate needs but can also ensure optimum performance and competitiveness in today's economic environment for any location. Maintenance excellence built on the true partnership with operations is the missing link to being the leader in the world market of tomorrow. A Site Leadership Team (opera-

FIGURE 18.1 Fitting the pieces together

tions, maintenance, stores/purchasing, and site functional groups as core members) drives activities to make maintenance a site issue. Maintenance must be viewed as an investment to add production capacity through more reliable equipment operation and workforce capacity through more efficient wrench time. Because of its positive impact on safety and quality, the focus on reliability is then given top priority.

The site's best practice that had the most positive impact was our operator and mechanic cross-training process. We created the concept of a shift maintenance coordinator (SMC) since skilled maintenance mechanics were in short supply in Singapore at the time. These SMCs were mechanics who had electrical, instrument, or mechanical (EIM) skills, and then we trained them in the remaining skills they lacked. This provided each shift with a maintenance leader with EIM skills. Because we were isolated on the island, our maintenance philosophy had to be that maintenance is a 24-hour-a-day job just like operations. Since we did not have mechanics to provide shift maintenance coverage, it was necessary to train the operators in the maintenance apprenticeship skills to the helper level as part of their orientation training. Each shift crew had one SMC who counted as part of the total operator allotment. These SMCs were experienced and trained to be able to provide full support knowledge for any maintenance situation. The

SMC reported directly to the operations shift supervisor and was part of the shift team. In 2015, the site had progressed to the point where the SMC was operationally crossed-trained and functioned as an operator on nights. This has contributed to the site being able to greatly reduce total head count and overtime as attrition occurred. Site staffing has been reduced from 160 to 105. The SMC is the maintenance mentor who leads by doing and showing. The shift supervisor decides which operators are assigned to help the SMC based on plant operations and area of knowledge. One operator per shift is assigned to serve as the backup for the SMC as part of the operator's development and performance progression.

Due to the shortage of skilled mechanics available to be hired, our plant decided to totally outsource our maintenance function but would only have coverage on days. Operators (supported by their SMC) are taught enough maintenance skills to allow them to handle emergency situations and make necessary repairs or equipment isolations to keep the plant in safe operation until permanent repairs can be performed by the contract mechanics on days. For shift repairs, the SMC will order and collect the needed parts from stores and plan the tasks to be done. Each shift has a crew toolbox maintained by the SMC. The SMC also maintains the computer equipment data for repairs made by his shift. The maintenance staff developed a list of routine and preventive tasks that the shift performs. History has proved that shift operators are closest to the equipment on a continuous basis and can prevent major failures by paying proper attention to warning signs and taking quick action. An understanding of maintenance allows the operators to better operate each piece of equipment and prevent potential damage by regulating its operating conditions. The quality and level of detail also increases for the work orders that they write to better describe the existing problem to the day mechanics.

A critical success factor, whether you outsource or do in-house maintenance, is to create a core reliability group. In this example, we had two mechanical engineers, one each for rotating and fixed equipment, and an electrical engineer for all instrumentation and electrical programs. As the owners, they analyzed the collected data from the operators and mechanics to determine the corrective actions to be taken. The three legs of the plant's reliability programs were preventive, predictive, and proactive.

- To prevent failures, we focused on lubrication and greasing, equipment alignment, calibrations and adjustments, motor current imbalance and insulation checks, replacement practices, and cleaning and housekeeping.
- The predictive processes were to detect potential failures and included vibration analysis, IR thermography, oil analysis and ferrography, non-

destructive technology, spike energy and cavitation, bearing temperature, corrosion monitoring, and operations checklists.

- The final phase was proactive, which focused on eliminating failures through the use of root cause failure analysis, laser alignment, standardization of equipment specifications, balancing of rotors, synthetic lubricants, criticality assessment, planning and scheduling, total productive maintenance practices, and outsourced maintenance with performance incentives. The rule for oil change decisions was five gallons or less change on a time basis and greater than five gallons change by analysis only.

Reliability, productivity, and capacity utilization are critical, especially since small changes in operating rates can make huge differences in earnings. With these types of programs and collected data, management will fund work and projects to support reliability and drive growth.

Development of troubleshooting skills by the shift personnel is required to identify and prevent small problems from becoming big ones. Shift operators take routine data trap vibration readings on nights to allow the maintenance staff to analyze the data for repairs by the contract mechanics. Preventive maintenance (PM) activities like maintaining proper oil levels are worked into the operator job routes. Operator maintenance efforts are directed by the SMC and focus on required jobs that could impact safety, production, or quality if not properly addressed. In small sites, teamwork is required among all parties (rotating shifts and day personnel) to maximize onstream time, product quality, and overall safe operations.

The results of implementing this work process allowed our plant to achieve an M&R (maintenance and reliability) cost as % ARV (asset replacement value) of 2.69% from our first year (1999) of operations compared with the company history for a new plant start-up average of 6.13%. In 2004, the site had progressed to 1.32%. The annualized maintenance costs during the 1999 start-up and construction was $6.28 million but went down to $2.2 million in 2003 and $1.85 million in 2004. The maintenance workforce head count went from 46 in 1998 for checkout and start-up to 28 in 1999 when the last unit went into operations and then 12 for normal operations in 2004.

Here is the performance for this plant in 2015 to show the long-term success:

- A capacity utilization above 90% achieved since start-up
- A YTD prime production above 99% achieved since start-up
- Excellent unplanned production loss control < 0.5% average per year

- M&R cost under budget since start-up with an average M&R cost of 1.5% ARV over the last 15 years (1.17% in 2013 as the best year)
- A 68% reduction of the maintenance contractor monthly average head count (FTE) since start-up with eight FTEs as the target
- Low maintenance overtime with 4.6% average total overtime and 0.8% average unplanned overtime
- High maintenance overall work schedule and PM schedule attainment above 97%
- Monthly work orders per FTE increasing from 38.5 in 2007–2009 to 60 in 2011–2013
- Work orders completed per year using the same number of craftspeople increasing from 3,611 in 2007–2009 to 5,188 in 2011–2013

This example can be used to create a great opportunity to obtain productivity gains as you develop improvement plans for your site. These concepts summarized above are expanded to provide additional details below as a maintenance vision, philosophy, strategy, and processes for you to consider for your application.

A Potential Future for Maintenance and Reliability

Maintenance Vision

Our maintenance strategy is data driven by our proactive preventive maintenance programs that utilize available predictive tools to identify and drive improvement. Our operators have adequate maintenance training to detect signs of initial failure and to take corrective actions to minimize equipment damage. A shift maintenance coordinator with electrical, instrument, and mechanical cross-training is part of the operations shift crew. Shift operator maintenance activities complement the day contract mechanic support through routine PM data collection for analysis, communication of operational impacts on equipment, and safe and proper preparation of units requiring maintenance. A shift SMC takes the place of one of the regular shift complement of operators and performs operating and maintenance tasks. Through his maintenance knowledge the SMC leads the other operators to handle equipment emergencies and repair decisions. This teamwork eliminates the need for shift maintenance coverage. Each shift has ownership for specific equipment and associated PM activities. SMC/operator maintenance experience allows the operators to troubleshoot and analyze problem causes and to be able to describe the situation to an expert if additional help is needed. For emergencies, they can understand and implement telephone instructions to eliminate imminent safety or equipment failure risks. The focus is to keep the SMC training up-to-date so that he can teach and coach his operators to grow their shift capability. SMC and operators utilize established maintenance processes and vendor documentation to solve problems needing immediate attention. They know their day maintenance contract resources and our staff capabilities to be able to handle problem escalations. Maintenance and operations utilize equipment criticality ratings and risk evaluations to minimize emergency work and to manage overtime.

The day maintenance forces consist of a skill-certified pool of mechanics with a high degree of cross-functionality. All specialty skills such as analyzer support have backup capability. Vendor relations are being continually refined to allow maintenance to optimize its capabilities while leveraging vendor expertise and resources. Day mechanics have an awareness and appreciation for maintenance costs and for the impact that their work has on production. All maintenance work is done safely or it is not done at all. There are close and supportive working relationships among operations, maintenance, stores, purchasing, HSE, technical support, and management to ensure obstacles are eliminated and continual improvement can be achieved.

The maintenance staff provides site support and consultation with other similar plants. The staff members frequently communicate with their counterparts to share experiences and to help each other solve problems. Best practices are identified and implemented at each site. Champions are selected to drive implementation efforts. Reliability engineers work with each site to leverage the resources and diagnostic capabilities of all locations to minimize duplicate expenditures by sharing expertise. A maintenance network provides flexibility and depth in skills available.

Our maintenance processes comply with and refine benchmarked worldwide maintenance and reliability guidelines. The team relationship among all similar sites is the key to our successful maintenance program. Our joint pride and ownership of the maintenance processes continues to grow through the confidence gained from understanding what needs to be done and why. This philosophy allows each individual to make meaningful contributions and to be recognized as a team player.

Maintenance Philosophy

Maintenance is a 24-hour-a-day job just like operations. Since we only have scheduled mechanics on days, it is necessary to train our operators in the maintenance (electrical, instrument, and mechanical) awareness skills as part of their overall skill development program. There is one SMC as a member of each shift crew. They are experienced mechanics who have worked in the electrical, instrument, or mechanical field in our industry. They report directly to the operations shift supervisor as part of his shift team. The SMC is the maintenance mentor who leads by doing. He will not do all the maintenance work performed on his shift but will guide the operators to handle assigned tasks. The shift supervisor will decide which operators to assign to help with maintenance based on plant operation and areas of knowledge. One operator per shift will serve as the backup for the SMC and will learn this job as part of the operator's development and performance progression. This assignment will be rotated over time to improve the overall coverage and skill set by each shift. The SMC will assist in plant operations when he has caught up on maintenance duties.

Operators will be taught enough maintenance skills to allow them to handle emergency situations and to make necessary repairs or equipment isolations to keep the plant in safe operation until permanent repairs can be performed by the day mechanics. Since true emergency callout mechanic support takes time to respond, the shift must be able to handle all situations that might occur. The SMC will evaluate each situation and determine if the work is within the scope of his shift personnel to complete, if it needs to be completed at this time, if it can be scheduled for a later time for his shift, or if it needs to be prepared for the day mechanics to do the next day. He will enter all work orders into the maintenance management system and do the necessary data input to specify needed parts and to fully define the scope of the required repairs if the work is passed on to the planner on days. If the work is to be done by the shift personnel, he will order and collect the needed parts from stores and plan out the tasks to be done. Each shift has a crew toolbox maintained by the SMC to be equipped to handle the tasks the crew is asked to do. Once repairs are made or tasks completed, he enters the history and his labor-hours to update the maintenance management system records. The maintenance reliability engineer develops a specific list of routine and preventive maintenance tasks to fit the skill set of the operators. These tasks help prepare the operators to handle emergencies and to prevent them from happening. History has proved that the operations personnel are closest to the equipment on a continuous basis and can prevent major failures by paying attention to warning signs and taking quick action. Vendor training is provided for critical items. An understanding of maintenance allows the operators to better operate each piece of equipment and prevents potential damage by regulating the operating conditions. The operators also create detailed work order scope descriptions that help communicate the exact needs instead of vague requests.

Development of troubleshooting skills by the shift personnel is required to identify and prevent small problems from becoming big ones. Our mission is to keep the site operating safely and efficiently. The SMC/operator maintenance work is not from the backlog but addresses ongoing needs that they identify. The operator maintenance efforts are directed by the SMC to focus on required jobs that could impact safety, production, or quality if not promptly addressed. This approach to maintenance is the future for all sites to drive productivity and cost control.

Maintenance Strategy

The majority of the maintenance work is performed on days by the maintenance mechanics. Maintenance provides a core group of mechanics daily through a contractor outsource partner. This allows mechanical and E&I (electrical and instrument) experts to be developed for each of our operating units and utilities. Shutdown coverage and swing load demand based on changes in the work order backlog will be met from a labor pool with contractor mechanics assigned to the appropriate core expert who will serve as a crew leader to guide repairs. The maintenance staff will troubleshoot and define job scopes and maintenance procedures with the assistance of the mechanic experts. They have established preventive and predictive programs for the shifts and day mechanics to implement.

All shift maintenance coverage is provided by the SMC and operators. The SMCs are the lead mechanics/coaches to guide the operators in making routine maintenance checks, performing program tasks, and making repairs required to continue operations until the day staff and mechanics can implement permanent solutions. The SMCs input their time into the maintenance management system and document equipment history. The general shift maintenance work is collected against standing orders, but specific equipment repairs are charged to individual work orders. The SMC does the planning for his operator crew and leads the job preparation for the day maintenance crews. Under SMC's leadership, shift operators do about 30% of the required routine maintenance. After consulting with his shift supervisor, the SMC makes the final decision to call out mechanics at night for emergency repairs when appropriate. SMC's are responsible for the daily communication of adequate details back to the planner to allow the efficient transition of work. The planners will also schedule activities for established PM programs into the daily work plans. They will enter day mechanic hours by work order and ensure that equipment history is captured.

The site field safety coordinators help enforce safety policies for all personnel. They implement our permitting process and perform audits to verify compliance. Lock-out tag-out and all other corporate and HSE guidelines are followed. First aid facilities are provided.

Day mechanics provide a turnkey service for all maintenance requests. Work orders in the backlog are managed to determine daily worker and prioritized work assignments. The planners coordinate all subcontracted special services (painting, insulation, etc.) and equipment needs. The appropriate third-party owner makes the initial contacts. We reserve the right to reject selected subcontractors based on performance or cost. Maintenance supervision works closely with the maintenance staff on required documentation and administrative functions. Major fabrication and machine shop services may be done at other facilities.

Maintenance Processes

A maintenance management system is utilized for all maintenance activities. Work orders are written based upon operations' needs on the shift that they occur. The SMCs enter each work order and open it if their shift team can do the task or put it into the backlog if it is to be done by the day mechanic crews. The planners provide plans to the scheduler for development of the work schedule for the maintenance mechanics. They interface daily with their supervisor to communicate expectations and determine completion dates based on priority and the equipment required to do the repairs. The reliability engineers will troubleshoot failures and implement improvements to increase reliability. They are responsible for the development of maintenance procedures as well as preventive and predictive processes. The planners enter and propose the schedule for PM tasks for the shift operators and day mechanics as appropriate. The engineers manage the maintenance programs such as painting, lubrication, and inspections. The administrative personnel ensure that labor-hours are properly entered into the maintenance management system and charged to the correct equipment locations. The maintenance manager is responsible for monthly invoice verification and approval for accounting to make payment.

For major shutdowns, the outage planners will develop the outage schedule based upon input from the operations shutdown logs, work order backlog, management of change records, annual inspections, and PM plans. This work will start several weeks prior to the shutdown to ensure the proper coordination of all the parties involved. Preplanning is also required to allow maintenance to secure the needed workers and other resources. The staff engineers will be assigned segments of the outage to manage including tie-ins and implementation of capital jobs.

Overall management of maintenance will be verified by performance measures. Monthly feedback sessions are held between the maintenance manager and her management.

Engineering assistance and/or implementation for small capital projects will be assigned to maintenance on a project-by-project basis. Major capital work will go through the engineering organization. The operations technical staff will manage this process with support from the maintenance staff.

Management of the spare part process is the responsibility of the maintenance manager. She works closely with the stores and maintenance staff to ensure data integrity. Working with stores, the maintenance manager establishes the minimum order point and order quantities. A free issue arrangement exists for high-use items.

This approach uses centralized planning with decentralized execution. We plan and schedule to improve how we do our work. A P&S best practice is to have the members of the shift crew (as they are leaving on their long weekend of 7-days off) to provide a list of jobs that they know needs to be done when they return on days. They can then be available with expert input and support as the jobs are performed. This also allows lead time for the planner to develop plans for the day mechanics.

Mechanic Skill Set

The following list of required skills is based on our experience following the processes outlined above but will need to be adjusted to fit your site:

Mechanical skills. The SMCs and day mechanical maintenance mechanics must be able to service and repair the following equipment and have the listed skills:

1. Pumps
2. Turbines
3. Compressors
4. Mechanical seals
5. Couplings
6. Alignment (laser and conventional)
7. Agitators
8. Filters and strainers
9. Shell and tube heat exchangers
10. Steam traps
11. Pipe fitting
12. Proper flange tightening
13. Tubing installation
14. Rigging

15. Pressure testing
16. Lubrication
17. Troubleshooting of operating problems
18. Troubleshooting of the root cause of failures
19. Computer skills
20. Print reading (vendor drawings)
21. Fabrication sketches
22. Verbal communication (English)
23. Written documentation

Electrical and instrument skills. Because of the complexity and level of computer control, the SMCs and day E&I mechanics must have knowledge in the following areas.

1. Electrical
   a. Theory
   b. Fusing
   c. Switches
   d. Relays
   e. Test equipment (multimeter, oscilloscope, clamp-on ammeter, megger)
   f. Wire sizing
   g. Print reading
   h. Hookup (basic electrical)
   i. Conduit running
   j. Starter/heater applications
   k. Hazardous area classifications
2. Analog electronics
   a. Control theory (closed loop, feedback, etc.)
   b. Total loop concepts
   c. Transmitters
   d. Controllers
   e. Calibration skills
   f. Voltage switch
   g. Current switch
   h. Resistors, diodes, silicon controller rectifier, semiconductor theory
   i. Transducers, converters
   j. Multipoint recorders

   k. Ohm's law
   l. Power supplies
   m. Recorders
   n. Soldering
   o. Thermocouples and RTDs
3. Digital electronics
   a. Programmable logic controllers
   b. Single-loop digital controllers
   c. Smart transmitter and handheld communicators
4. Pneumatics
   a. Concepts
   b. Control valves (troubleshooting and stroking)
   c. Tubing
   d. Transmitters
   e. Positioners
   f. Tuning
   g. Pressure switches
5. Distributed control systems (DCS)
   a. DCS concepts
   b. Console operation
   c. Input/output measuring
6. High voltage/power
   a. Safety
   b. Breakers
   c. Transformers
   d. Fuses
   e. Battery maintenance
   f. Uninterruptible power supply function
7. Analyzer
   a. Combustible gas detectors
   b. Calibration
8. Other
   a. Lock-out tag-out
   b. Math
   c. Logical troubleshooting
   d. Computer skills

e. Verbal communications (English)
f. Written documentation

Operator Shift Maintenance Duties

The following duties are conducted by the SMC/operators as part of their normal routine shift activities:

1. Reset the 230-volt circuit breakers.
2. Check the electrical fuses using a fluke.
3. Reset the 400-volt motor starters.
4. Take amperage readings for troubleshooting.
5. Block in and zero the transmitters.
6. Verify the operation of a control valve and make adjustments.
7. Perform instrument calibrations.
8. Conduct instrument and electrical troubleshooting.
9. Run tubing.
10. Connect and disconnect small motors.
11. Calibrate a flow meter to verify proper operations.
12. Isolate equipment (complete the LOTO process).
13. Change out faulty gauges.
14. Switch gas cylinders.
15. Properly tighten the flanges.
16. Install and remove the blind flanges.
17. Troubleshoot the analyzer systems and unplug the lines.
18. Add packing to a valve.
19. Prepare the exchangers for cleaning.
20. Properly lubricate the equipment (oil and grease) and take samples.
21. Take routine vibration readings.
22. Conduct field inspections and document the findings for corrosion, vibration, and damage (insulation, painting, structural, leaks, etc.).
23. Troubleshoot the rotating equipment (speed, seal systems, unusual noises, etc.) and make needed corrections.
24. Perform specified preventive maintenance tasks.
25. Change out the steam traps.
26. Unplug the process lines (NPG, etc.).
27. Change out the filter elements.

28. Troubleshoot the vacuum systems and make corrections.
29. Hook up a portable air compressor and use pneumatic tools.
30. Use a chain fall.
31. Make up hose connections.
32. Change out the small equipment (pump, motor, etc.) for emergency repairs.
33. Do a rough alignment of the equipment train to have it ready for emergency operations.
34. Adjust or change out a regulator.
35. Do alarm checks and verifications.
36. Zero and change bad CGA heads.
37. Make short conduit and wire runs.
38. Replace the small valves.
39. Repair small process, water, and air leaks.
40. Open the vessels for cleaning.

Once these tasks are achieved, maintenance will be in tune with the site needs and will proactively initiate corrections and enhancements. Maintenance will be an established well-coordinated effort with all parties having an active role in work identification, priority setting, and completion of tasks. Through proper proactive systems using the appropriate predictive technologies, reliability is improved to ensure that equipment is available and in optimum condition to perform whenever needed. Reliability will reveal the "hidden" plant for extra capacity with minimum capital expenditures.

This maintenance transformation will give a competitive advantage to all industries that choose to implement it. It has been proved to finally break the insanity cycle and propel your organization into the future you have always wanted to achieve.

I have included several files and tools at www.maintenanceinsanity.com to help with your journey of discovery and success. Please contact me at RDLSolutions@Outlook.com for additional information or support to ensure you never have to worry about maintenance insanity again.

The "Maintenance Insanity" Cure Support Files

The following files and tools are available at www.maintenanceinsanity.com to help you apply the processes and cures described in this book:

- Six Bucket Cost Analysis Tracking Tool
- Planning and Scheduling Training Slides for Class Exercise
- Productivity Implementation Checklist and Action Items
- Schedule Performance Tool with Metrics
- Contractor Productivity Workshop Kickoff Presentation
- Fish Scoreboard for Crew Teams
- CAIP Charts and Process Tool
- Maintenance Assessment Worksheet Blank
- Deep Dive Template and Agenda for Contractor and Owner Site Visit
- Service Provider Milestones Transition Plan for Client Facility
- Site Transition Plan
- Master Turnaround Checklist
- Maintenance Process Assessment 21 Questions
- Maintenance Morning Report
- Typical Interview Questions
- Sample Contractor Crew Improvement Projects
- Shutdown Planning and Scheduling Guidelines
- Daily Maintenance Planning, Scheduling & Execution
- Planning Decision Tree
- Small Site Planning and Scheduling Slides
- Sales & Operations Planning Organization
- Continuous Alignment Improvement Process Tiers with Initial and Six-Month Score

Please contact me at RDLSolutions@Outlook.com for additional information or support to ensure you never have to worry about maintenance insanity again.

Index

References to figures are in italics.

www.ingramcontent.com/pod-product-compliance
Ingram Content Group UK Ltd.
Pitfield, Milton Keynes, MK11 3LW, UK
UKHW051911100926
8556IPUK00005B/852

9 780831 136246